T0285711

Side Effects

Side Effects

How Left-Brain Right-Brain Differences Shape Everyday Behaviour

LORIN J. ELIAS, PH.D.

DUNDURN
PRESS

Copyright © Lorin J. Elias, 2022

All rights reserved. No part of this publication may be reproduced, stored in a retrieval system, or transmitted in any form or by any means, electronic, mechanical, photocopying, recording, or otherwise (except for brief passages for purpose of review) without the prior permission of Dundurn Press. Permission to photocopy should be requested from Access Copyright.

Publisher: Scott Fraser | Acquiring editor: Russell Smith
Cover designer: David Drummond
Cover image: Shutterstock.com/Jolygon

Library and Archives Canada Cataloguing in Publication

Title: Side effects : how left-brain right-brain differences shape everyday behaviour / Lorin J. Elias, PH.D.
Names: Elias, Lorin J., 1972- author.
Description: Includes bibliographical references and index.
Identifiers: Canadiana (print) 20220199922 | Canadiana (ebook) 20220199973 | ISBN 9781459747555 (softcover) | ISBN 9781459747562 (PDF) | ISBN 9781459747579 (EPUB)
Subjects: LCSH: Cerebral dominance. | LCSH: Brain—Localization of functions. | LCSH: Left and right (Psychology) | LCSH: Human behavior.
Classification: LCC QP385.5 .E45 2022 | DDC 612.8/25—dc23

We acknowledge the support of the Canada Council for the Arts and the Ontario Arts Council for our publishing program. We also acknowledge the financial support of the Government of Ontario, through the Ontario Book Publishing Tax Credit and Ontario Creates, and the Government of Canada.

Care has been taken to trace the ownership of copyright material used in this book. The author and the publisher welcome any information enabling them to rectify any references or credits in subsequent editions.

The publisher is not responsible for websites or their content unless they are owned by the publisher.

Printed and bound in Canada.

Dundurn Press
1382 Queen Street East
Toronto, Ontario, Canada M4L 1C9
dundurn.com, @dundurnpress 🐦 f 📷

To Lana, the best left-hander

Contents

Introduction

I think of my body as a side effect of my mind.
— CARRIE FISHER

Human behaviour is lopsided. Our bodies are relatively symmetrical, at least on the outside, but the way we behave is not. Chances are, your left hand and right hand don't look much different from each other, yet almost 90 percent of us prefer to use the right for writing, throwing, and almost any activity requiring considerable skill. However, when cradling a newborn child, most of us cradle the infant to the left. When posing for a portrait, whether hand-painted in the 16th century or a modern selfie on Instagram, we tend to put our left cheek forward. When kissing a lover, we tend to tilt our head to the right. Why is our behaviour so lopsided and what does this teach us about our brains? How can we use this information to make our selfies more attractive when posting them to our online dating profiles, or how can we Photoshop our political advertisements to make the candidate more appealing to a particular political demographic? Can knowing how left-brain right-brain differences shape our opinions, tendencies, and attitudes help us make better choices in art, architecture, advertising, or even athletics? By the end of this book, you will see how lateral biases in our brains influence our everyday behaviour and how you can use this information to your advantage.

The functional differences between the two sides of the brain are easy to detect in a scientific laboratory, in a hospital's brain scanner, or from an individual's behaviour after a one-sided brain injury or brain surgery. However, these left-right differences are also readily observable in normal people as they simply go about their daily routines. Our lopsided behaviour is hidden in plain sight.

Some of our left-right differences are very strong, consistent, and *old*. For example, 90 percent of us favour the right hand, and it doesn't matter whether you are male or female, from Malaysia or France. Furthermore, our species has been right-hand dominant for more than 50 centuries, according to analyses of ancient artifacts and artworks.[1] Other strong lateral preferences are more recent, only hundreds of years old, such as posing biases in portraits. If we take a careful look at religious artworks depicting the Crucifixion of Jesus, we find that 90 percent of them depict Jesus turning his head to the right and putting his left cheek forward.[2] Our lateral posing biases for cradling babies are also quite old, but they're nowhere near as strong — about 70 percent leftward — as with our population-level hand preferences.

The lopsided behaviour surveyed throughout this book relates to our underlying left-right brain differences. Everyone's brain is unique. Just like the unique face each person displays on the outside of their head, most everyone has the same parts inside their skull, with the same general shape, location, and function. But each brain is unique. All of the left-right differences encountered in this book are *at the population level*. In other words, the biases discussed here are trends that apply to a large group of people but not necessarily to every individual within the group. Take handedness, for instance. There is a population-level bias toward right-handedness, but some people are left-handed. We know that 90 percent of people are right-handed[3] but that doesn't mean there is anything wrong or fundamentally different with each left-hander in the world. The same holds true for other individual differences in brain lateralization. The left side of the brain is dominant for language in 90 percent of us,[4] but that doesn't mean the 10 percent with right hemispheric language are any less proficient with the spoken or written word.

In some cases, these individual differences in lopsided behaviour can be revealing. For example, most new mothers cradle their infants to the left, but rightward cradling is more common among mothers with depression.[5] If you prefer to cradle your little one to the right, does that mean you are depressed? Absolutely not. However, in a population of people who prefer rightward cradling, depression is more common *in the group* compared to those who cradle to the left. This book is an exploration of population-level and group trends, not individual diagnoses or analyses.

While I am issuing disclaimers, one more note of caution is necessary before we proceed. The left-right differences surveyed here are *relative*, not absolute. I'm a member of the right-handed majority, but my left hand is not completely useless. I routinely employ it for relatively unskilled tasks, such as picking up and carrying objects, and I can use my left hand to shoot a basketball well enough to win the occasional game of H-O-R-S-E. Similarly, I know from a functional magnetic resonance imaging (fMRI) scan that the left hemisphere of my brain is dominant for language. However, that does not make my right hemisphere functionally illiterate. It can read and understand words in several languages, but not with the same speed, fluidity, and nuance — particularly when word order is important — as my left hemisphere. As I said, the differences between the hemispheres are relative, not absolute. Even if the differences *were* absolute — and they are not! the two hemispheres of the brain are interconnected with more than 250 million projections from one side to the other by an impressive white-matter structure called the corpus callosum.[6]

The two hemispheres of the brain collaborate to form perceptions, memories, and even biases. Stating that one half or the other is solely responsible for any of those things is more than an oversimplification; it is usually wrong. Consider this example: my teenage daughter, Mileva, walks into the living room, and I exclaim, "Nice shoes!" Her left hemisphere is probably more proficient at decoding spoken words, and on its own, it might interpret my statement as a compliment. However, what if I used a *sarcastic* tone when saying those two words? "*Niiiiice* shoes!" The pitch and tone decoding expertise typically dominated by the right hemisphere of the brain would detect that shift in meaning, and Mileva could then scoff at her father's lack of fashion sense as she leaves the room.

Disclaimers aside, there are many structural, chemical, and functional differences at the population level between the two halves of the brain. The right hemisphere tends to be larger and heavier and contains more white matter (brain cells covered in a fatty insulation called myelin) than the left and is more diffusely organized and interconnected.[7] In comparison, the left hemisphere is smaller and denser and contains a higher proportion of grey matter (brain cells). If we take a typical brain out of the skull and look at it from above, it tends to exhibit a counter-clockwise torque,[8,9] with the frontal lobe of the right hemisphere extending farther forward and the occipital (rearmost) portion of the left hemisphere extending farther back (see Fig. 1). There are also many more lateral differences within the hemispheres, such as the planum temporale[10] (a structure associated with language processing) normally being larger in the left hemisphere. These are just the physical differences. This book is really about the *functional* differences.

The most well-known functional difference is left hemisphere dominance for language. We know from studies of brain injuries, brain surgeries, and functional brain imaging that the left brain dominates language for 90 percent of us. It also excels at perceiving word order (i.e., syntax) to make meaning of phrases (consider "dog bites man" versus "man bites dog"), perceiving rhythm in music, performing logical ordering, and planning sequences of movements. Conversely, the right hemisphere excels at perceiving emotion (especially negative emotion), spatial information, pitch/melody in music, facial recognition, tone of voice in speech, and perceiving themes in pictures, sounds, and spaces.[11]

One of the most striking and puzzling features of the vertebrate nervous system is its *contralateral organization*. For every vertebrate animal known (even *Agnathans*, jawless fish from the Cambrian period), but no known invertebrates, the right side of the brain controls the left side of the body, and the left side of the brain controls the right side of the body.[12] The same reversal holds true for information coming *into* the nervous system. A touch on the left hand is perceived by the right hemisphere of the brain and vice versa. For some of our senses, this crossover is more complete than for others.[13] For visual information, almost all of it on the left side (not through the left

4

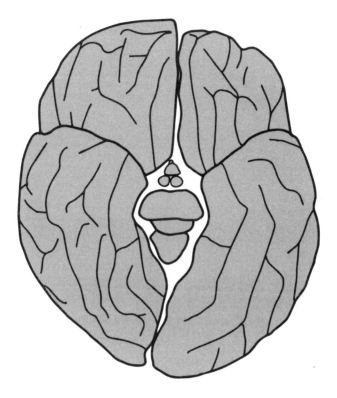

Fig. 1: In this view of the brain from below, the right frontal lobe normally extends farther forward and the left occipital lobe extends farther to the back of the brain, giving the brain a torqued, counter-clockwise appearance when viewed from above.

eye, but the data from the left side going to *both* eyes) projects to the right hemisphere (see Fig. 2). Hearing is a little different, with about 70 percent of the information from the left ear projecting to the right hemisphere of the brain.[14] Why is our nervous system crossed over? I have no idea. It just is.

As you read this book, you are going to get your lefts and rights confused at times, but that won't be your fault. It doesn't mean there is something wrong with you. I will require some fancy mental gymnastics from you at times, and with the help of a picture or two, I am confident that we can navigate these lefts and rights together. Fig. 2, which shows the crossover in

The Visual Projection Pathway

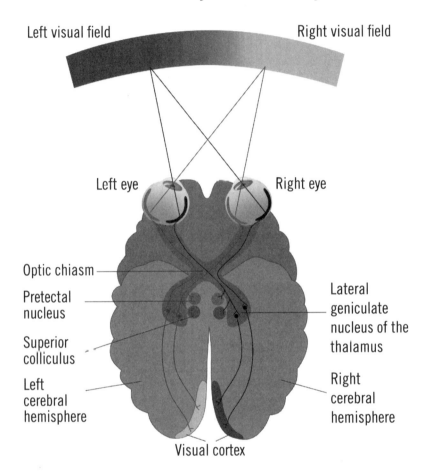

Left visual field

Right visual field

Left eye

Right eye

Optic chiasm

Pretectal nucleus

Superior colliculus

Left cerebral hemisphere

Lateral geniculate nucleus of the thalamus

Right cerebral hemisphere

Visual cortex

Fig. 2: The human nervous system exhibits contralateral organization in which information coming from one side of space is primarily processed by the opposite side of the brain. This figure details the crossover for visual information, but the same principle applies for most other sensory systems and the control of movement.

the visual system, seems simple enough: left space goes to right brain and vice versa. However, when I start to describe how the right hemisphere's specialization for facial recognition is responsible for the left-cheek bias for selfies taken in the mirror, you have to imagine a person's face centred in the visual field, imagine which half of the face is in which space for two people facing each other, and then reverse your lefts and rights yet again in your mind because the scenario being discussed is someone looking at him or herself in the mirror!

This book is organized in a way that presents our lateral biases one at a time, which can give the impression that they are always independent of one another, but they are not. For example, hand preferences (see Chapter 1) are strongly related to our lateral preferences for feet, ears, and eyes (see Chapter 2). The posing biases in portraiture (see Chapter 6) are also related to the lighting biases (see Chapter 7) we glimpse in the same pieces of artwork. This doesn't mean that one bias necessarily *causes* the other, but just because the biases are discussed in separate chapters doesn't mean they are discrete, independent phenomena. Many of them connect to one another, and the Afterword is dedicated to stringing those together once we have examined each bias on its own.

1

Handedness: Are Left-Handers Always Right?

I drink coffee with my right hand, and I smoke
with my left. But I talk with both hands.
— GEORGE BURNS

The most famous and obvious side effect of all is handedness. Our lop-sided brains are plain to see through our universal, population-level preference for the right hand. Handedness is not a new discovery or a new development. It is even mentioned in ancient texts such as the Bible. But handedness somehow manages to be the most studied, yet most mysterious left-right difference of them all. It's the one lateral preference everyone has noticed, and even if you haven't read one of the dozens of books devoted exclusively to handedness, I would guess that you have reflected on your own hand preferences at one point or another. Even a minor injury to the dominant hand serves as a cogent, even embarrassing reminder of how useless that *other* hand can be.

For all these reasons, an analysis of handedness is both the best and the worst chapter to open our discussions of the lefts and rights of everyday life. But I don't really have a choice. As you will discover in every subsequent chapter, handedness influences, not causes but influences, most of the other lateral biases examined in this book. I couldn't escape the complicated and omnipresent impact of handedness when considering the other lateral preferences even if I wanted to. In keeping with the focus of all the other chapters, I will restrict my discussion of handedness to the everyday-life lens through which all the other lateral biases arc portrayed.

What proportion of the population is right-handed? I can offer short and long answers. The short one is that approximately 90 percent of us are right-handed. But the long answer? The proportion depends on where someone is born. It depends on *when* the person was born. It depends on the culture and environment in which the person was raised. It depends on the sexual orientation of the person. It even depends on the developmental trajectory of the person and whether something might have gone awry during the birth process or even before. It depends on a person's gender. But it only depends *so much*. These factors might push that approximate 10 percent figure around a bit, but only just. Outside of a left-handers convention — though there are virtual and even some in-person gatherings for left-handers, especially on August 13 (International Left Handers Day) — you will never find a point in time or space or culture where more than 50 percent of a large group prefers the left hand.

Let's start with time frames. Modern image archives are rich with data about handedness. We can review imagery such as photos of people signing important documents, or even look on the back of a baseball card from decades ago, to surreptitiously assess the handedness of an individual long after they have died. But these types of records only go so far back. If we ask *how long have humans been mostly right-handed*, where does that answer lie? Early written records are scarce. For example, Judges 20:15–16 in the Bible describes a battle whose participants consisted of 700 left-handed or ambidextrous men and 26,000 right-handed combatants. Even this proportion (97 percent) heavily favours the right-handed.

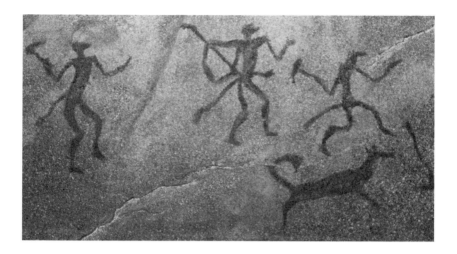

Fig. 3: Cave drawings from time periods spanning centuries tend to depict people holding items in their right hands.

However, we can go back even farther than the Bible, perhaps even millions of years. The hunting style of Australopithecus shows signs of right-handedness![1] Paleolithic stone tools reveal strong evidence of rotation of the stone core by right-handed toolmakers.[2] A similar pattern is seen in stone implements constructed by Peking Man.[3] Furthermore, nearly every early culture on the planet produced images of people engaged in various activities, such as hunting. In some of these pictures (see Fig. 3, for instance), we can clearly observe a person using one hand instead of the other to throw or carry an object. A strong bias toward right-handedness can be seen in paintings of hands by Cro-Magnon people,[4] examination of North American Indigenous art,[5] and the hand employed for skilled activities depicted in paintings in the tombs of Beni Hasan and Thebes between 2500 and 1500 BCE.[6]

In a survey of more than 12,000 artworks[7] that clearly portray an action with one hand or the other, drawn from as early as 15,000 BCE up to 1950 CE, 92.6 percent preferentially feature the right hand. This bias was remarkably stable over time; for example, it was 90 percent for images from before

Fig. 4: Paintings of mostly left hands from a cave (Cueva de las Manos) in Patagonia, Argentina. The images were probably made by the ancestors of the Aónikenk people around 700 CE. Despite the fact that most of the images depict the left hand (829 left hands compared to 31 right hands), these images are considered evidence of right-handedness because the left hand presumably served as a "stencil" while the right hand held a spray pipe made out of bird bone to create the image.

3000 BCE and between 89 and 94 percent from 500 to 1700 CE. Using this admittedly unusual research technique, it appears as though the occurrence of right-handedness has been largely unchanged for the past 50 centuries![8]

But the story isn't quite that simple. About 3 percent of people born close to 1900 CE were left-handed, but before and after that time, levels of left-handedness were around the 50-century average of 10 percent. One of the best handedness datasets in the world came about mostly by accident. When I was a child in 1986, my parents were faithful *National Geographic* subscribers, and in September of that year, a most unusual version of the magazine arrived in our mailbox. It had a "scratch-and-sniff" card inside, and more than 11 million subscribers were asked to identify the smells they could from the card, answer a few demographic questions, and mail in the card. Two of those demographic questions were about handedness, asking for both the writing and throwing hands. The response was incredible. More than 1.4 million people returned the completed cards, and even though handedness did not appear to be linked to olfactory discrimination — one of the questions motivating the survey in the first place — the interrelation of the demographic variables was extremely revealing, especially in a dataset that large.

The original report from this dataset pointed out two key findings. One was interesting and the other one was puzzling and somewhat alarming. The interesting finding was that men were about 25 percent more likely than women to report being left-handed. It's the second finding that has drawn most of the attention. Left-handedness is relatively common among respondents born after 1950 (remember that the survey was in 1986, so someone born *after* 1950 would be 36 or younger), but left-handedness was less and less common among earlier born, older people, plummeting to only 3 or 4 percent of respondents born on or before 1920. A second, smaller dataset covering the same time period — this time sampling mostly British instead of American citizens — replicated the same pattern.[9] Where were all the left-handers born between the late 1800s and 1920? Was left-handedness so much less frequent among people from this cohort, or worse, could left-handers be suffering higher rates of mortality, resulting in very small proportions of elderly left-handers?

The simplest and scariest explanation is the latter one. Perhaps left-handers tend to die earlier than right-handers. On the surface, this seems like a really easy hypothesis to test. All you need to do is look at mortality/life-expectancy statistics for left-handers versus right-handers. However, the intricacies in designing and *interpreting* the results of some recent studies might surprise you. Before we get into specific studies, I need to detail two very different approaches to problems like this. When psychologists study changes over an individual's lifespan, they can do so in two very different ways. A cross-sectional study takes place all at once and surveys people of different ages, at the same time. The *National Geographic* study I described earlier did exactly that. Longitudinal studies are different. Those follow the same individuals over a period of time, testing them over and over, searching for changes across the time span.

Cross-sectional studies of handedness and age are very clear. They consistently find high rates of left-handedness (usually over 10 percent) among young people and very low rates of left-handedness (often 2 to 5 percent) in older adults.[10,11,12] There are plenty of potential explanations for this effect. The most obvious one is social pressure. A century ago, the social pressure against left-handedness was enormous. My own relatives have shared harrowing stories about left-handed classmates having their left hands tied behind their backs, forcing them to write right-handed. Practices like that were meant to convert natural left-handers to "forced" right-handers (In Chapter 12, Sports, you will also learn about some examples of the reverse phenomenon — encouraging athletes in particular sports to develop as forced left-handers). Perhaps large numbers of left-handers are not disappearing at all. Maybe they are simply being forced to convert to right-handedness. An even simpler version of this "social-pressure" explanation is that the pressure results in a decreased willingness to *report* being left-handed.[13]

However, the story does not end there. There are also studies that follow the same individuals over a period of time, and these can also detail decreased longevity among left-handers. Research of this type is rather difficult. You need to have accurate and complete records that include both handedness and mortality measures/dates for a large group of people. Where would you go looking for such records? You get a gold star if you guessed "sports" and platinum if you said "baseball" records in particular. Baseball is a notoriously statistics-obsessed sport, and among those records are details that include the handedness of the player.

Using statistics from *The Baseball Encyclopedia*,[14] Stanley Coren and Diane Halpern retrieved the data about every player detailed in the book that passed away before 1975, then calculated an individual's risk of death at a given age, taking their handedness into account. They started at age 20 and found that left-handers were no more likely to die that early compared to the right-handed players. However, beginning at age 33, they found that left-handers were about 2 percent more likely to die before their next birthday than right-handers. Two percent might not sound like a lot, but year after year it really adds up, and it alone can account for the relative scarcity of left-handers among the elderly. The difference in average lifespan of right-handed players compared to left-handed players in the encyclopedia was less dramatic — 64.64 years for right-handers compared to 63.97 for left-handed players.

This study received a tremendous amount of media attention, including a feature in *Weight Watchers Magazine*, but it also drew a lot of criticism. Some of the most enthusiastic critics were themselves left-handed, which seems a conflict of interest. However, these left-handers raised several excellent criticisms. In addition to the concerns about the statistical methodology in the original study — details I won't reiterate here — critics also raised the excellent point that only certain positions in baseball tend to draw left-handed players. To be more specific, the left-handed players are often pitchers. Pitching also happens to be a particularly stressful job on a baseball field. What if it's *stress* that leads to an early death, not handedness at all? The contention that left-handers have shorter lifespans is still contentious today. Coren and Halpern have replicated and extended their lifespan effect,[15,16] and a similar result has been observed in cricket players.[17] However, a number of other studies have failed to find a similar effect,[18–21] and still others have criticized the methodology employed in these studies.[22,23]

We've established right-handedness as the norm across time but have not considered the question of *regional* variation. The prevalence of left-handedness does indeed vary depending on location, but the variation we see is a matter of *degree*, not direction. There are no left-handed cultures. This sentiment was expressed perfectly in 1836 by Thomas Watson, an English physician, in the *London Medical Gazette*:

> The employment of the right hand in preference to the left is universal throughout all nations and countries. I believe no people or tribe of left-handed persons has ever been known to exist.... Among the isolated tribes of North America which have the most recently become known to the civilized world, no exception to the general rule has been met with.

Watson's assertion from almost 200 years ago is still just as true today as it was then, but it doesn't address the question of regional variation in the rate of left-handedness. A consensus is building that such variation exists, but there is no agreement on the rates themselves, partly because of methodological differences between samples from different countries and partly because the sample sizes from some nations are very small, making comparisons to very large samples from large countries rather difficult. Another complication is that linking the nation of residence to ethnicity or culture is an increasingly challenging generalization to make.

There are many studies that simply compare the rates of left-handedness across two countries, and some general themes emerge from those pairings. For example, one study found a rate of left-handedness of 9.8 percent in Canada but only 4.7 percent in Japan.[24] A similar study compared rates in Canada versus India, yielding a parallel difference. The prevalence of left-handedness in India was only 5.2 percent.[25] Surveys of left-handedness in Asian countries generally report very low rates, often in the range of 3 to 6 percent, whereas Caucasian samples produce estimates of left-handedness rates that are double, even triple.[26]

One method of trying to separate the racial/ethnic contribution to geographic differences from the genetic ones is to look at migrated populations. In a 1998 study of thousands of medical school applicants in the United States, 13.1 percent of the Caucasian applicants were left-handed compared to 10.7 percent of Black applicants, 10.5 percent of Hispanic applicants, but only 6.3 percent of Vietnamese applicants, 5.4 percent of Korean applicants, and 5.3 percent of the Chinese applicants. Similar ethnic distributions are evident within the very large American dataset from *National Geographic*.[27] Collectively, these results suggest an underlying genetic influence, not simply a cultural one.

Handedness runs in families. This claim is not controversial. Handedness is heavily influenced by the environment a person is raised in. This claim is also not contentious. However, reconciling those two claims, especially when they can produce opposing predictions, *is* highly controversial. When reading this section about the genetics of handedness, please keep the following disclaimer in mind: there is no simple answer here, and just because there's overwhelming evidence of a genetic contribution to handedness, that does not necessarily mean that a single gene controls the behaviour.

According to a meta-analysis[28] of studies of patterns of handedness in families (a meta-analysis statistically combines the results of many different studies), two right-handed parents have a 9.5 percent chance of having a left-handed child. However, the likelihood rises to 19.5 percent if only one parent is left-handed. Curiously, this one-parent effect appears to be stronger on the mother's side of the family. If both parents are left-handed, the chances of a child being left-handed rise to 26.1 percent. Considered by themselves, these statistics do not necessitate a genetic explanation. After all, lots of non-genetic things run in families, such as recipes for cinnamon buns or an affection for Swedish automobiles. Effects like these can be driven entirely by parental pressure. However, when adoption studies are thrown into the mix, there's pretty convincing evidence in favour of a genetic mechanism. The handedness of an adopted child is more likely to follow that of his or her biological parents than that of an adopted parent.[29,30]

If we think back to the Mendelian genetics lessons we enjoyed in high school, the numbers that might have impressed in the previous paragraph suddenly stop resonating. Remember filling out those Punnett squares during a final exam? It didn't matter if we were predicting the colour of a pea in a pod or the odds of a parent having cystic fibrosis, the only percentages that mattered when calculating alleles with those Punnett squares were 0 percent, 25 percent, 50 percent, and 100 percent. Left-handedness proportions of 7 and 21 percent didn't fit. Squinting hard enough, we might have been able to blur 26 percent into 25 percent, but even then, getting odds of left-handedness up that high requires *both* parents to be left-handed. Clearly, handedness runs in families but doesn't appear to be under the simple and predictable control of a single dominant or recessive gene.

Despite these inconvenient mathematical proportions, most early genetic theories of handedness proposed that handedness is indeed a recessive trait and that its proportions should follow the laws of Mendelian genetics.[31] However, the complexity of the inheritance pattern of left-handedness does not conform to this type of genetic model. Most recent genetic theories of handedness account for this complexity by either proposing a poly-genetic model involving more than one gene or add an element of "chance" to a single-gene model. For example, Chris McManus and M.P. Bryden[32] proposed a genetic theory where one allele, D, codes for right-handedness, whereas the other allele, C, corresponds to chance and can result in either right- or left-handedness at equal proportions. Therefore, a DD individual will be right-handed 100 percent of the time, 75 percent of CD or DC individuals will be right-handed, and CC homozygotic individuals will have a 50/50 chance of developing right- or left-handedness. Even these proportions fail to match the prevalence data perfectly, although they come much closer than the earlier, simpler theories.

Regardless of whether there is one gene or many, or whether there is an element of chance, it's extremely unlikely that genes code for handedness itself. Instead, genes probably code for another process or substrate that influences handedness, and the genes themselves must interact with the environment to be activated in the first place. It's apparent that handedness is influenced by genes, but it's just as clear that these influences are complicated and probably interact with the individual's environment, including his or her culture. Next time a flashy headline is spotted such as the BBC's "Left-Handed DNA Found" from September 5, 2019, don't be fooled into thinking there is a simple solution to this particular riddle.

It's evident that handedness runs in families, but it's also obvious that genes are only part of the reason why. What else might lead to left-handedness? Well, the environment, of course. Our genes don't exist in a vacuum, and it's the environment that determines when and how and in what combination genes express themselves. The environment can trigger reactions, even restrict their range or suppress them. The influence of the environment could also affect handedness more directly instead of modulating the effects of our genes. These environmental influences can include social and parental pressures, especially early in life. They can also involve the intrauterine environment, with mechanical factors such as fetal position, stress during birth, or chemical influences like hormones (although the latter is likely influenced by genetics, as well) coming into play. They could even arise from abnormal cell division and twinning.

Perhaps the simplest of these explanations is the claim that parental pressure shapes the handedness of a child. After all, we have already established that the more left-handed parents (and other relatives) children have, the more likely they are to develop left-handedness themselves. John Jackson,[33] founder and honorary secretary of the Ambidextral Culture Society, was an early proponent of this view. In 1905, he claimed that most humans are right-handed because their parents were right-handed and that any given child can be right-handed, left-handed, or even ambidextrous when raised in the appropriate environment. To take advantage of this flexibility, Jackson suggested that all children should be taught to use either hand interchangeably, making them functionally ambidextrous. Abram Blau, head of child psychiatry at New York City's Mount Sinai Hospital in the 1940s,[34] also credited parental influences for the subsequent handedness in children, but he took a rather more negative view, influenced by Sigmund Freud's psychodynamic perspective. Specifically, Blau asserted that left-handedness was usually the result of "emotional negativism" in early childhood and that handedness has no biological basis whatsoever.

Strictly environmental explanations like these have some obvious shortcomings. Handedness runs in families, and it's entirely possible that *some* of that effect is due to environmental influences. However, handedness occurs in biological families independent of the handedness and presumably the environment of the people actually parenting the child. In adoption studies, the handedness of a child is more closely related to that of the biological parent than that of the adoptive parent.[35,36] We also know that population-level left-handedness has been relatively stable for 50 centuries now.[37] If the environment alone determines handedness, and that environment is typically negative (sometimes even violently so) against left-handedness, why would the trait persist across centuries? Furthermore, even when the environment and the genes for two individuals are virtually identical, such as in the case of identical/monozygotic twins raised together, those two individuals often don't exhibit the same handedness.[38,39] Finally, we see evidence of hand preference in utero[40] long before any postnatal influences have the opportunity to determine the handedness of the child.

Others have taken a more anatomical approach, crediting handedness to some of our other obvious and reliable left-right physical differences. After all, the left-right differences we see in the gross anatomy of the brain pale in comparison to the anatomical asymmetries evident elsewhere in the human body. Perhaps the most obvious and well-known example is the leftward displacement of the heart. However, even some of our paired organs, such as the lungs and kidneys, or sex organs like the ovaries and testes, exhibit very obvious and reliable anatomical asymmetries.[41] Might these left-right differences in our bodies lead to population-level biases toward right-handedness?

Perhaps the most famous of these claims is the sword-and-shield theory, often credited to British historian and essayist Thomas Carlyle (1795–1881), although Lauren Harris[42] has identified physician and medical scientist Philip Henry Pye-Smith (1839–1914) as an even earlier source. Harris quotes Pye-Smith as follows:

> If a hundred of our ambidextrous ancestors made the step in civilization of inventing a shield, we may suppose that half would carry it on the right arm and fight with the left, the other half on the left and fight with the right. The latter would certainly, in the long run, escape mortal wounds better than the former, and thus a race of men who fought with the right hand would gradually be developed by a process of natural selection.[43]

Charles Darwin's newly published theory of evolution would have been front of mind for the Victorian scientists of the day. As such, the sword-and-shield theory would have gained traction because of its ingenuity and simplicity. However, this theory predicts the emergence of population-level right-handedness *after* the Bronze Age, when swords and shields first appeared. We know from cave drawings[44] and prehistoric tools[45] that right-handedness was the norm *long* before the appearance of the first sword. Furthermore, the early combatants described in this theory were typically male, so one would predict *lower* rates of left-handedness in males because of the sex-specific selection pressure. However, there are five male left-handers for every four female left-handers,[46] a small but very reliable sex difference in the opposite direction predicted by the sword-and-shield theory. Finally, there are extremely rare cases of situs inversus, in which asymmetries, including the heart and other organs, are reversed from left to right.[47,48] Individuals with situs inversus do not exhibit left-handedness any more frequently than individuals without the unusual condition. In a sample of 160 people with the rare anomaly, only 6.9 percent were left-handed.[49]

Another "cardiac" theory will be explored much more thoroughly in Chapter 5 when we investigate images of parents cradling their children. Leaving those representations aside for the moment, the central claim of the parent-holding-baby theory is that parents tend to hold infants to the left (toward the heart), and this leftward-holding arrangement helps soothe the child with the sound of a parental heartbeat but also leaves the parent's right hand free to perform skilled/complex tasks while child-rearing.[50] Evidence supporting this theory can be found almost anywhere, including centuries-old paintings of parents and children, or minutes-old Instagram posts of parents and their children. Parents clearly prefer to cradle to the left.[51] In addition to accounting for this bias, the theory also correctly predicts the greater prevalence of right-handedness among females. Given that females have traditionally performed more of the child rearing duties in most cultures, the pressure toward right-handedness should be stronger if it's driven by leftward cradling biases.

However, as an explanation of the cause of handedness this theory falls short in several important ways. Perhaps the most obvious flaw is that the theory focuses on the handedness of the parent instead of the child. By the time humans reach child-rearing, or even child-holding, age, handedness is already firmly established. Furthermore, a leftward cradling arrangement should have more influence on handedness development in the child, not the parent. When cradled to the left, the child's right hand is "pinned" next to the body of the parent, but the left hand is relatively free for reaching and grasping movements. If this holding bias influences handedness at all, it should encourage left-handedness in the child. Despite these flaws, the cradling bias, as we'll discover in Chapter 5, is consistent across gender, time, geographical location, and even species.

Other developmental theories of handedness are more controversial, and some are even disturbing. The most extreme version of this perspective posits that left-handedness itself is the result of something gone awry during development, such as brain damage during the act of birth itself. Presenting left-handedness as a "syndrome"[52] attracted a lot of attention. In the late 1980s, an eye-catching explanation emerged from neurologists Norman Geschwind and Albert Galaburda.[53] They described their theory as "triadic,"[54] but it was popularly known as the prenatal-testosterone theory for good reason. It claims that elevated levels of prenatal testosterone are responsible for deviations from the brain's "normal dominance pattern." The theory was appealing for several reasons. One of them was the convincing and charismatic way Geschwind himself advanced the theory. Another was the simple and elegant mechanism proposed. But the third was the way this theory seemed to account for a large number of previously unrelated and inexplicable correlations with handedness.

Perhaps the most obvious of these correlations is the sex difference in handedness. Left-handedness is more common in males, making the assertion that elevated testosterone is involved intuitively sensible. The theory is also consistent with the higher prevalence of autoimmune and language disorders among males, the different maturational rates of the two sexes, and the association between left-handedness and allergies, autoimmune disease, cerebral palsy, Crohn's disease, dyslexia, eczema, Rett syndrome, schizophrenia, and thyroid disease.[55] Elevated prenatal testosterone levels help explain these correlations because testosterone can influence the growth of many tissues and has an inhibitory effect on the growth of immune structures such as the thymus gland. Testosterone can also influence the development of several brain structures, including specific nuclei in the hypothalamus and limbic system.

Despite its appealing simplicity and ability to account for some previously puzzling relationships between handedness and several pathologies, the position that an elevated level of prenatal testosterone results in left-handedness leaves us with more questions than answers. Why does testosterone selectively slow growth of the left hemisphere of the brain and not also the right hemisphere? If we measure testosterone levels in amniotic fluid before a baby is born, then follow up with that same person 10 to 15 years later to assess handedness, why are higher testosterone levels linked to higher rates of *right*-handedness,[56] not *left*-handedness? Some of the associations between left-handedness and pathological conditions have been difficult to replicate.

Claiming that left-handedness itself is pathological is not new. Abram Blau presented his emotional negativism theory, which is arguably another variant of that controversial position. More recently than Blau, some have claimed that left-handedness typically occurs due to stress on the brain during birth.[57] A less extreme variant maintains that left-handedness is *sometimes* the result of pathology,[58] or at the very least it serves as a marker for other pathologies. The evidence for this is many of the same associations between left-handedness and disorders already discussed earlier, as well as associations with other disorders, such as ulcerative colitis, stuttering, skeletal malformations, psychoticism, post-traumatic stress disorder, myasthenia gravis, migraines, epilepsy, deafness, and coronary artery disease.[59] Perhaps the best evidence in favour of the theory is the connection between left-handedness and reports of birth stress as seen in low Apgar scores.[60] (An Apgar score is a scoring system out of 10 meant to assess health and well-being in newborn infants, taking into account breathing, heart rate, muscle tone, reflexes, and skin tone.) Left-handedness is also linked to premature birth[61] and low birth weight.[62] All these traits are also more common in twins, and we already know that left-handedness is more widespread in them. In fact, once you control for the twinning effect, some of these associations with left-handedness disappear.[63]

There are other problems with the proposal that left-handedness results from birth stress or other pathologies. Despite major technological advances in health care in general, and supports for the birth process in particular, the prevalence of left-handedness hasn't gone down. If it's changing at all, the frequency of left-handedness is on the rise. The theory would also predict higher rates of left-handedness in countries with less sophisticated health care, but there is no evidence for those trends, either. Finally, it might be true that left-handers are more numerous among certain medically vulnerable groups, but left-handers are also overrepresented among the "gifted," including individuals with abnormally high IQs, musicians, architects, lawyers, children of professionals, students of the visual arts, and the intellectually precocious.[64] There is even some evidence that left-handed men earn more than right-handers with similar levels of education.[65]

No discussion of potential causes of left-handedness is complete without describing the disturbing and unforgettable "vanishing-twins" theory. Some have called it the "undercover-twins" theory, which is a little less disturbing, but the underlying details are the same. We have already established that handedness runs in families, that twinning is also prevalent in families, and that left-handedness is more common in twins. But there's more. A relatively small number (15 to 22 percent) of monozygotic (identical) twins exhibit "mirror imaging" of various physical features such as hair whorls, fingerprints, or birthmarks. Most case reports focus on mirror imaging of physical anomalies, especially dental ones. However, there are even some very rare case reports of *complete* mirror imaging, with one twin exhibiting situs inversus.[66] Since you are reading a chapter about handedness, it should come as no surprise that this mirror imaging in monozygotic twins can also apply to handedness. In these cases, one twin develops as a right-hander, whereas the other becomes a left-hander.

So we know that left-handedness is more common in twins, and some twins exhibit mirror imaging, including handedness, but what's this *vanishing* business? Well, when a pregnant woman exhibits multiple gestations during an ultrasound, one (or more) of the gestations might not survive. As early as 1976, Salvator Levi reported that a shocking 71 percent of the multiple gestations observed during an initial ultrasound *disappeared* and that the pregnancy ended with the birth of a singleton.[67] Since then, others have reported disappearance rates between 43 to 78 percent.[68] As shocking as this might be for most of us, some reproductive scientists are not particularly surprised. According to C.E. Boklage, "The loss of one member of a twin pair can be understood quite simply as part of the highly imperfect biology of human reproduction. Most human conceptions fail before birth. It is no different and no more mysterious for twins."[69] But here's the mysterious part: the vanishing twins theory posits that left-handers typically have a vanished twin, and presumably 10 percent or so of right-handers do, too. That's unsettling. Perhaps many of us, not just left-handers, are survivors of our vanished twin.

However, it's very unlikely that the vanishing-twins theory can account for all, or even many, cases of left-handedness. Even if the mechanism proposed by the theory is accurate, the math doesn't add up. Let's assume that left-handed fetuses are just as viable as right-handed fetuses, and for every left-hander that is the survivor of a right-handed, vanished twin, there should also be a right-handed survivor of a left-handed vanished twin. Approximately 10 percent of the North American population is left-handed. So, even if every monozygotic twin pair exhibited mirror imaging, 20 percent (10 percent for left-handers, 10 percent for right-handers) of all pregnancies would need to have multiple gestations at one point or another to account for the current prevalence of left-handedness. However, multiple gestations are only present about 3 percent of the time. Furthermore, only 15 percent of the twins that survive full term exhibit mirror imaging. Allowing for this value, the prevalence of multiple gestations would have to be greater than 100 percent to account for all left-handers.

To finish off how and when left-handers are born, let's consider the time of year. Surprisingly, a number of studies have detected elevated rates in the birth of left-handers during particular seasons. In one study of almost 40,000 people,[70] left-handers tended to be more common among births between March and July, but only in the northern hemisphere and only among males. This effect was reversed in the southern hemisphere. Other large studies have failed to find the same effect.[71,72] Still other surveys indicated that male left-handers tended to be born during fall or winter,[73] mostly aligning with another study showing that left-handers were more frequent among births between November and January.[74] However, an examination of half a million people in a UK Biobank failed to reveal this effect among the males in the dataset but did discover a very small elevation in the frequency of left-handedness among women born during the summer months.[75] Collectively, these results are confusing and point out that very large sample sizes are needed to detect some very small effects, if they exist at all. It's also possible that any seasonal effects are limited to specific parts of the world (perhaps locations with large climate differences between the seasons), particularly those with warmer summers.[76]

For the left-handed reader, this paragraph about biases in manufactured goods, especially tools/implements, will contain few surprises. It's no secret that left-handers live in a right-handed world and that our population-level bias of 90/10 percent translates to even stronger biases among manufactured goods. Scissors are the obvious example, but that one is so evident that left-handed scissors are relatively easy to find, at least to cut paper or fabric. More specialized shears for leather, metal, pruning of plants, or even cutting hair are much more difficult to obtain. The vast majority of our kitchen implements are designed for right-handed operation, including can and bottle openers, soup ladles, peelers, and measuring cups. Left-handers can use right-handed versions of these items, but it's inconvenient, uncomfortable, and can even result in minor injuries. But things get worse. Industrial tools are even more likely to be designed for the right-hander, and attempting left-handed operation of a right-handed drill press, band saw, table saw, or jointer can be very dangerous. Unsurprisingly, reports of left-handers being more accident-prone than right-handers are often credited to the machinations left-handers need to go through to use right-handed implements.[77]

TAKEAWAYS

Our current population-level bias toward right-handedness is nothing new. It is clear that across time and around the world right-handedness is the norm. There are definitely variations in the relative prevalence of right-handedness, but there is no evidence to suggest there was ever a time or place when left-handers outnumbered right-handers. It is also obvious that handedness runs in families but that simple genetic explanations for this effect can't account for every left-hander. Some developmental and environmental mechanisms contribute to handedness, and this might help explain the high incidence of left-handers among certain groups. Based on our current high rates of left-handedness among young people and low frequency of left-handedness among the elderly, it can be tempting to conclude that left-handers don't live as long as right-handers. However, factors such as social pressure are responsible for at least some of that disparity. As we will see in subsequent chapters, many of the other side effects are influenced by handedness, although it is unlikely that any of them are singularly caused by differences in hand preference.

2

Feet, Eyes, Ears, Noses: Starting on the Right Foot

There is as much expression in the feet as in the hands.

— NICOLAS CHAMFORT

Sidedness is a myth. The idea that a person generally prefers one side of the body for almost everything (i.e., consistent preference for left hand, left foot, left eye, left ear, et cetera) is clearly wrong. Yes, there are some people with very consistent lateral preferences across the whole body, but they are few and far between. In some cases, that kind of extreme and consistent sidedness is actually the result of a developmental or acquired disorder. It's a sign that something is wrong. Instead, being somewhat "mixed" is a lot more normal. In the previous chapter, we discovered that approximately 90 percent of us are right-handed. Lateral preferences for *almost everything* other than handedness are weaker than that. Foot preference — footedness — has a bias of three-fourths or four-fifths to the right. Ear and eye preferences are biased by about two-thirds to the right. So even some basic knowledge of math immediately tells us that many people will have mixed preferences. If 90 percent are right-handed, 80 percent are right-footed, and 66 percent are right-eyed, it means that *at most* two-thirds of people have consistent preferences for those three things and only for rightward preferences. Having a right bias for one thing doesn't necessarily predict any others, although lateral biases might give clues about the other ways the brain is functionally lopsided, i.e., which side is dominant for language.

As we discovered in the previous chapter, handedness can be influenced by many factors, and biology is just one of them. Despite overwhelming evidence that handedness is influenced by our genes and that the development of hand preference starts *before* birth, our environment, including culture, can have profound and permanent impacts on hand preference. Cultural practices can even dictate which hand is used for "clean" or "dirty" activities. However, cultural influences for other side preferences are very difficult to discern. If we scour religious texts for guidance about which eye or ear or nostril to use in a particular situation, that information is scarce indeed. It's also quite easy to uncover instances where left-handed individuals were coerced or forced to switch from left-handed writing or eating to using the right hand, but similar instances focusing on feet, eyes, and ears are extremely rare. Given this lack of interference, might these other asymmetries provide a more "pure," culturally unadulterated glimpse into an individual's natural lateral preference?

FOOTEDNESS

A preference for one foot or the other is probably quite strong, maybe not as powerful as hand preference, but that is not a particularly fair comparison. How often during a typical day do we interact with an object with feet? We probably don't spend much time picking things up with feet or writing our names in the sand using big toes. Instead, any real "footwork" is probably accomplished during tasks such as driving or playing a sport like soccer, although, ironically, probably not participating in American football, almost all of which is done with hands. Compare that to how many times a day we pick something up or move an object with our hands and it probably comes as no surprise that feet are an afterthought.

However, around 80 percent of us prefer the right foot.[1] There are plenty of ways to measure that. When I was a graduate student, I developed a short questionnaire for assessing footedness (see Fig. 5). It asks two kinds of foot preference questions. The first type is about using the foot to manipulate an object (kicking a ball, picking up a marble, smoothing sand at the beach), whereas the other types of questions focus on maintaining posture or support.

FIG. 5: QUESTIONNAIRE FOR ASSESSING FOOT PREFERENCE

Instructions: Please indicate your foot preference for the following activities by circling the appropriate response. If you **always** (i.e., 95 percent or more of the time) use one foot to perform the described activity, circle **Ra** or **La** (for **right always** or **left always**). If you **usually** (i.e., about 75 percent of the time) use one foot, circle **Ru** or **Lu** as appropriate. If you use both feet **equally often** (i.e., you use each hand about 50 percent of the time), circle **Eq**. Please do not simply circle one answer for all questions, but imagine yourself performing each activity in turn, then mark the appropriate answer.

1. Which foot would you use to kick a stationary ball at a target straight ahead?   **La Lu Eq Ru Ra**

2. If you had to stand on one foot, which foot would it be?   **La Lu Eq Ru Ra**

3. Which foot would you use to smooth sand at the beach?   **La Lu Eq Ru Ra**

4. If you had to step up onto a chair, which foot would   **La Lu Eq Ru Ra**
    you place on the chair first?
5. Which foot would you use to stomp on a fast-   **La Lu Eq Ru Ra**
    moving bug?
6. If you were to balance on one foot on a railway   **La Lu Eq Ru Ra**
    track, which foot would you use?
7. If you wanted to pick up a marble with your toes,   **La Lu Eq Ru Ra**
    which foot would you use?
8. If you had to hop on one foot, which foot would you   **La Lu Eq Ru Ra**
    use?
9. Which foot would you use to help push a shovel into   **La Lu Eq Ru Ra**
    the ground?
10. During relaxed standing, most people have one leg   **La Lu Eq Ru Ra**
    fully extended for support and the other slightly
    bent. Which leg do you have fully extended first?
11. Is there any reason (i.e., injury) why you've changed **Yes No**
    your foot preference for any of the above activities?
12. Have you ever been given special training or   **Yes No**
    encouragement to use a particular foot for certain
    activities?
13. If you have answered Yes for either question 11 or 12, please explain:

_____

_____

_____

We have learned a lot about handedness by conducting studies of data collected by sports teams where handedness data is routinely collected (baseball, cricket, et cetera). Similarly, many studies of footedness come from the world of sport, especially soccer. This soccer data is complicated by the fact that players are coached from a very young age to develop both feet as equally as possible.[2] Indeed, a soccer player who is too one-footed is often regarded as the product of poor coaching.

By most estimates, 80 percent of the general population is right-footed.[3-7] Similar to the handedness data, very young people show higher rates of leftward foot preferences, whereas people over 60 show very high rates of right-foot preference. Because foot preference isn't subject to the same cultural pressures as is handedness, it might actually be a more "pure" measure of laterality for movement. Some studies, including some of my own, have found that footedness is a better predictor of cerebral lateralization than handedness is.[8,9] In other words, knowing whether you are right- or left-footed makes it easier to predict the side of the brain that dominates language processing, compared to trying to make that same prediction based on handedness. This is a surprising and counterintuitive finding. After all, we engage our hands in communication all the time, both through the written word and by gesturing during speech (see more about gestures in Chapter 9).

EYEDNESS

Most of us are fortunate enough to have two fully functional eyes. Unless we are dressing up as a pirate for a Halloween costume party and patching one eye for fun, we typically enjoy two very similar but distinct views of the world at the same time. Because of the small distance (five to seven centimetres or so) between the two pupils, each eye conveys a slightly different view of the three-dimensional world, and one of the more clever tricks played by the visual system is to use the disparities between those two pictures to help us perceive depth. The differences between these two images — collectively referred to as *binocular disparity* — give the brain depth cues, and the bigger the differences, the closer the objects are (see Fig. 6).

Most of the time, we take in the visual world through both eyes. However, there are clearly times when one eye dominates vision. If we look through a telescope, peep through a keyhole, examine a specimen with a monocular microscope, or gaze down the sight of a rifle, we tend to use our

Fig. 6: Because of the distance between the two eyes, the images processed by each differ slightly, an effect exemplified in this illustration attributed to Peter Paul Rubens from Franciscus Aguilon's *Opticorum* (1613).

dominant eye. For two-thirds of us, that's the right one. Since we already know that almost 90 percent of the human population is right-handed, we can clearly see that handedness and eyedness don't go together. Sure, most right-handers are also right-eyed, but most left-handers are also right-eyed.

Scientists have been writing about eye dominance for at least 400 years, starting with the Italian polymath Giambattista della Porta in 1593,[10] but relatively little is known about it, especially compared to handedness. There are as many as 25 different ways to measure eye dominance, from asking people which eye they use to look through a telescope to inviting them to demonstrate eyedness through a monocular task. In my own lab, I ask people to put both hands together as if praying, leaving a small hole between them, then prompt them to look at my nose through the hole between the hands. Next, I record the eye that appears through the hole (see Fig. 7).

Fig. 7: You can assess eyedness by asking someone to put both hands together in a configuration resembling praying and asking the person to view your nose through the small hole between the hands. By employing both hands equally, this controls for the possibility of the person simply raising the dominant hand in front of the eye on the same side.

In a very large meta-analysis of other studies of eyedness, including 54,087 participants,[11] the prevalence of right-eyedness is almost exactly two-thirds of the population. Eyedness was more concordant with handedness when both were measured with questionnaires instead of performance measures, but that could be the result of "contamination" with the questionnaire method. Some people simply select the same type of answer for each question. (See the sample questionnaire in Fig. 5 and imagine a participant simply selecting "right always" or "right usually" for all items instead of actively considering the scenario posed in each question.)

That same huge analysis revealed something unusual about eyedness. It confirmed that eyedness is loosely associated with handedness, but one of the most interesting findings was something they didn't find — sex differences. For almost every lopsided behaviour, males demonstrate the effect more prominently or more often than do women. For example, males are more likely to be left-handed than women are. Even with the huge sample size of almost 55,000 people, there was no difference between the men and women in terms of eyedness. Weird.

Like handedness, eyedness appears to run in families. The frequency of left-eyedness increases continuously with the number of left-eyed parents, but the pattern of inheritance does not follow any straightforward dominant Mendelian model,[12] nor does eyedness appear to be as heritable a trait as handedness.

EAREDNESS

Ear preference shares many commonalities with eye partiality. Most of us benefit from two functional ears, and the brain can use the differences in loudness and even arrival time of sound to help perceive the location of a sound's source. There are certainly times when we present a single ear to hear something, such as when we put an ear to a door to eavesdrop on a conversation or raise a telephone to one ear. Like eyedness, two-thirds of us are right-eared, and ear preference does not relate to handedness in a simple way.[13,14] Most right-handers are right-eared, but most left-handers are also right-eared.

NOSTRIL PREFERENCE (NOSEDNESS?)

It's easy to imagine how preference for one hand or foot can influence our behaviour. After all, when signing a name or kicking a ball, we only need one limb. It's also easy to articulate an individual limb, right or left. Athletes and musicians can learn to exert exceptional independence in these movements on both sides of the body at the same time. It's a little harder to discuss eye preference or ear inclination, not just because we need to focus on sensation instead of movement but because we normally see with both eyes and hear with both ears. To really delve into a discussion of eye or ear dominance, we need to invoke some special circumstances such as when we look into a monocular microscope with one eye or hold a phone against one ear. The nose is yet another paired sensory organ. Most people have two nostrils, and of course all the wondrous and even much less wondrous odours of the world typically permeate both of those nostrils at roughly the same time with the same intensity.

However, each nostril has its own sensory pathway to the brain,[15] and quite unlike the projections from our eyes or ears, the dominant projections from the nose are ipsilateral; they primarily feed the hemisphere on the same side as the nostril. Stimulation of the left nostril primarily affects the left hemisphere and vice versa for stimulation of the right nostril.[16] This is also weird. Sensory projections from our hands, feet, eyes, and most every other body part are processed on the opposite side of the brain from the original source of the stimulation. Not so with the nose. Given these pathways, is it possible that our two nostrils have differential abilities/sensitivities and that we systematically prefer one or the other, just as we do for other sensory organs?

There are a few different ways we can study left-right sensory differences with our noses. First, we can consider the sensitivity of one nostril or the other to detect a particular scent. We do this kind of thing all the time, as when we walk into the kitchen and wonder if we smell gas from the stove, or perhaps when we sniff a drink to see if there is alcohol in it. That kind of sensitivity is for the detection of a particular smell. Is it there or not? A second way we can examine the act of smelling is by looking at our ability to discriminate between one scent and another. We do this when comparing perfumes or colognes or when choosing flowers for a bouquet. A third and very different way of examining our nasal behaviour is to consider changes in airflow and resultant alterations in how we feel and think. Most of us have a lot of experience with the first two scenarios, but the third is a lot less common.

Let's start with the first method — examining the sensitivity of one nostril or the other. There are only a few studies available, and they have yielded some rather mixed results. Anecdotally, perfumers often find that one nostril is more sensitive than the other.[17] One study looked at the differences in smelling sensitivity for n-butanol — an alcohol with a pungent smell — between the two nostrils in 19 males (regular people, not perfumers). It found that right-handers were more sensitive with the right nostril, whereas left-handers were more receptive with the left.[18] However, other studies have found that left-handers were more responsive with the right nostril,[19] and still other investigations have failed to find any relationship between handedness and sensitivity for odour detection between the two nostrils.[20] Nevertheless, for odour discrimination (differentiating one smell from another), a very different pattern of results emerged. Whereas Robert J. Zatorre and Marilyn Jones-Gotman[21] found that both right- and left-handers were better at odour discrimination with the right nostril, Thomas Hummel and colleagues[22] observed that right-handers discriminated between odours better with the right nostril, whereas left-handers were more effective with the left.

Now, to complicate matters further, the two nostrils typically engage in a pattern of spontaneous and reciprocal congestion/decongestion. Under normal circumstances (i.e., not when you have a cold) when one nostril clears up, the other constricts, and vice versa. This was first detected and recorded more than a hundred years ago.[23] We now use the term *nasal cycle* to refer to the swelling of one nostril and the shrinking of the other.[24] Between 70 to 80 percent of us experience this regular cycle, and a hot wire anemometer can actually be used to measure the relative airflow of each nostril. Alan Searleman and colleagues[25] did this and found that when people were asked which nostril was passing more air, they often guessed incorrectly. However, they measured more leftward airflow in left-handers and more rightward airflow in right-handers.

Because of this cyclical pattern, it's possible that the hemisphere getting more air is more activated during that portion of the cycle. Several studies have found increases in either verbal or spatial processing, depending on whether the left (verbal) or right (spatial) hemisphere was getting more airflow during that portion of the nasal cycle.[26,27]

One of the strangest studies I ever conducted myself was an examination of the effects of forced single-nostril breathing on cognitive performance.[28] First, we determined the "dominant" nostril by having our participants breathe through their noses on mirrors. The side producing the bigger "cloud" was the dominant one at the moment, exhibiting more airflow. Then we had people either breathe through their dominant or non-dominant nostril while performing listening tasks. Our participants had to identify the emotional tone of voice for the rhyming words *bower, dower, power,* or *tower* spoken in a tone of voice that was either happy, sad, angry, or neutral. Participants who were right-nostril dominant and were forced to inhale through the dominant (right) nostril had strong right hemisphere (left ear) advantages for detecting the emotional targets. Therefore, the one-sided breathing appeared to enhance the activation of the right hemisphere for the task. So the next time you plan to try something that will really tax the right hemisphere of your brain — perhaps remembering your way through a corn maze or composing music — consider some right-sided forced nostril breathing first.

Outside of that niche application, alternate-nostril breathing is a technique embraced by many yoga practitioners and was first described approximately 5,000 years ago. Yoga has demonstrable impacts on memory, and there is even some evidence that alternate nostril breathing enhances non-verbal memory, such as the recall of numbers or memory for spatial locations.[29]

TAKEAWAYS

Our strong preferences for one side or another are not limited to our hands. We also tend to exhibit strong and reliable lateral biases for our feet, eyes, ears, and even nostrils. Furthermore, these preferences don't tend to "match" across organs. Ten percent of the population favours the left hand, but 30 percent prefers the left eye. Most left-eyed people are right-handed. In those rare instances when people have very consistent and strong lateral partiality across the whole body, this can signal the presence of a developmental or acquired disorder. Despite the typical lack of congruity across these lateral inclinations, foot, eye, ear, and nostril favouring can alter our perceptions and actions, and in some cases (like unilateral nostril breathing), we can even leverage these preferences to our advantage.

3

Words: The Left Isn't Treated Right

To the right go honours, flattering designations, prerogatives: it acts, orders, and takes. The left hand, on the contrary, is despised and reduced to the role of a humble auxiliary: by itself it can do nothing; it helps, it supports. It holds.

— ROBERT HERTZ[1]

How many times have we heard the phrase "No, your other left"? Keeping our lefts and rights right can be tricky at the best of times. If we suffer any left-right confusion at all, writing a book like this one is a minefield. For all the effort I've put into checking, double-checking, and triple-checking my lefts and rights as I worked through each topic, I was genuinely fearful about accidentally getting a left or right wrong in one fateful paragraph, eventually leaving a careful reader literally in left field. But I am getting ahead of myself. We will discuss the positive and negative biases associated with lefts and rights shortly.

When teaching people how to keep their lefts and rights straight, I often resort to the trick of having people hold up both hands, extending the thumb and index finger of each hand, tucking the other fingers out of the way. Only one hand makes a proper L when that's done — the left hand. The right hand makes a backward L. Of course, people who have mirror-reversal issues with letters won't be helped with this tactic, either. More permanent solutions are also available such as the pair of modest tattoos in Fig. 8.

Making left-right confusion errors can be especially problematic in medicine, and as seen in an example from a famous television medical drama, *House, M.D.*, some patients even resort to writing on their healthier body parts before surgery to make sure the correct limb is targeted. My son recently had his third foot surgery, and the fastidious surgeon signed the targeted spot on the right foot during the pre-anaesthesia consult, just to be certain.

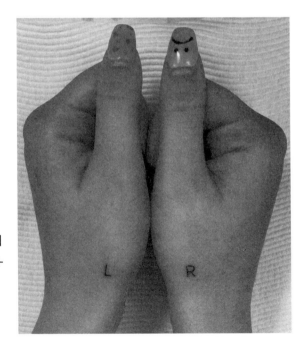

Fig. 8: A surefire and permanent method to differentiate the left and right hands, posted by Australian tattoo artist Lauren Winzer after a client requested this navigational aide.

Of all the asymmetries we face in the world, left-right differences are often less dramatic than up-down or front-back ones. As we will see in Chapter 7, our assumptions about the source of light include a very strong one about light coming from above but a weaker one about light also emanating from the left. As Alan Cienki, professor of language use and cognition at Vrije University in Amsterdam, states, the "left-right spatial axis is very weakly polarized in terms of the human body and our daily conscious functioning: it plays a much less significant role than do the up-down and front-back axes."[2,3] Perhaps that is why we so readily learn up/down (by two years of age) and front/back (by four years of age) as children, but many struggle to differentiate left/right even well into adulthood.[4]

If keeping your lefts and rights straight is a challenge, you aren't alone. Around one in five of the general, neurologically typical population report *frequent* confusion of left and right orientations, and women are more than twice as likely to cite the condition. I'm not talking about trouble remembering whether the correct route to the grocery store ends in a left or right turn, or if the right or left electrical switch turns on the kitchen lights or the garbage disposal. Keeping those kinds of lefts and rights straight is clearly impossible for everyone — right? I am talking about the failure to correctly *identify* left and right.

Many otherwise perfectly intelligent people have trouble differentiating between left and right. Terrifyingly, one of the most popular studies of the phenomenon was conducted testing medical students. Those researchers found that over 15 percent of the students struggled with lefts and rights. Let's hope that 15 percent didn't tend to choose to specialize in surgery. Some of the most famous medical errors of all time are "wrong-sided surgeries," resulting in the incorrect organ being removed or treatment of the wrong limb. In one case, two doctors accidentally removed the lone functional (left) kidney instead of the right one, a mistake that proved fatal for the patient.[5]

Among a sample of almost 500 people, 15 percent of the participants were terrible at correctly identifying the side of the indicated hand using the test shown in Fig. 9, and almost half employed a hand-based (acting-out-the-scenarios) strategy for solving the puzzles.[6] That might have been hilarious to watch; however, left-right confusion is no joke. It leads to

L   R   R

Fig. 9: A test of left-right orientation in which participants need to identify the side of the shaded hand, modelled after a study by Sonja Ofte and Kenneth Hughdahl (see Chapter 3, note 5).

navigational errors, motor vehicle accidents, surgical mistakes, and incorrect signs, advertisements, and even typos in books like this one.

Now let's complicate matters further. Throughout our discussion of left-right confusion, the same simple and consistent words have been applied to all things right and left. However, there are a great many other descriptors within English and other languages to describe lefts and rights, many of which are value-laden. We will also discover that this is not an equal opportunity endeavour. Not only are the words for the left rather negative, even derogatory, but there are a great many more of them than the positive terminology we reserve for the right.

Many words differ in valence, and left/right are no different. Although some words are relatively neutral, such as *street* or *spoon,* emotional interjections such as *yuck* or *wow* have explicit emotional valences, and other words have more subtle, implied ones, such as *spider* or *party.*[7] Some people have claimed that the words *left* and *right* are in the neutral category, but by the end of this chapter, I hope to convince readers that across languages and cultures, the implied valence for words meaning "left" is typically very negative, whereas the implied valence for words meaning "right" is typically positive.

Even the meaning of rightward or leftward movement conveys this bias. For the Māori of New Zealand, the right side is the side of life and strength, whereas the left is the side of death and weakness. Some Indigenous North Americans use the right hand to stand for *me*, whereas the left refers to *the other*. A raised right hand signifies "bravery, power, virility; while on the contrary the same hand, turned to the left and placed below the left hand, signifies, according to context, the ideas of death, destruction, and burial."[8] In many religions, a happy place must be entered stepping first with the right foot, and the presentation of offerings must be performed with the right hand. When sinners are expelled from a church, they leave by the left door. In funerary rites, and in the case of exorcisms, the ceremonial circuit is made in the opposite direction as normal (leftward instead of rightward).

In religious ceremonies around the globe, the left hand is typically despised, even humiliated. When it actually wields a power of some sort, "the power of the left hand is always somewhat occult and illegitimate; it inspires terror and repulsion."[9]

In some tribes of lower Nigeria, women are forbidden to use their left hands while cooking. Apparently, this is partially to avoid being accused of attempts at sorcery or even poisoning during food preparation, since both those tasks would presumably be accomplished with the left hand, as French sociologist Robert Hertz (1881–1915) maintained.[10] In Arabic cultures, there is also a hygienic dimension to the left-right divide. While the right hand is employed for eating, drinking, and food preparation, the left is reserved for cleaning oneself.[11,12]

The religious symbolism of right and left is also reflected in its imagery. In Western art, when Eve accepts the apple from the devil, resulting in expulsion from Paradise, she takes the fruit with her left hand. Mayan rulers are depicted holding objects in their right hands, and subordinates are pictured to the left. Defeated enemies might even be depicted as left-handed.

In Christian texts such as the Credo, Jesus sits at the right hand of God. In Buddhism, there are two paths to achieving wisdom: the rightward for those following social conventions and adhering to ethical codes; the leftward for those espousing the breaking of taboos and the abandonment of morality.[13]

The words we use to describe left and right are similarly biased against the left. Many of the left-right distinctions in Indo-European and even

Fig. 10: Figurine of a Mayan ruler (Prince of Diamonds) holding a fish in his right hand.

non-Indo-European languages are based on splits between opposing source words such as *straight/curved, strong/weak, clean/dirty, male/female, high/ low, older/junior, leader/follower,* or *light/dark.*[14]

This dichotomy has resulted in a long list of derogatory terms referring to the left. For example, the old English term for *left* — *lyft* — originally meant "lame" or "weak." In Gaelic, *cli* has the negative connotations of being "unhandy," "unpractical." Some of the words from the Bantu family of languages for *left* have to do with things that are "forgotten, dried up" or even "crooked, horned."

Some of the gendered terms have interesting origins. In Bakongo, the right hand is *kooko kwalubakala,* "the hand of men." Conversely, the left hand is *kooko kwalukento,* "the hand of women." In New Guinea, the right side is *sidik tam,* meaning "good" or "proper," but the left is *kwanim tam,* which is derived from the verb *kwanib,* meaning "to roll." One common chore for women in New Guinea was to make bags by rolling fibres. They would typically perform this act with the left hand against the left thigh while feeding new fibres through with the right hand.[15]

Some languages base the terms for *left* and *right* on other directions, perhaps north-south or east-west. In Sanskrit, *dakhsina* means both "right" and "south." In Arabic, *shamaal* means both "north" and "left." In both Biblical Hebrew and Classical Arabic, "south" and "right" are denoted with *yamiin*.[16]

Conversely, our terms for *right* demonstrate a clear bias toward the correct, true, and active. In Russian, words for *right* correspond to "straight" or "correct." These also refer to the norms in the society, which are also considered to be positive. Following an analysis of the words for *left* and *right* in 50 languages, medical anthropologist Wulf Schiefenhövel[17] identified the source meanings for words referring to *left* and *right*. The source meanings for *right* were positive terms such as *straight, strong, clean, higher, leader,* and *light* compared to the leftward source meanings such as *curved, weak, dirty, follower,* and *dark.*

In addition to all these "literal" left-right biases in languages around the world, there are also more figurative partialities in our expressions. A "left-handed opinion" is weak or wrong. A mistress is a "left-handed wife." A bad dream is left-handed. A "left-handed compliment" is an insult. In Dutch, if you neglect something or someone, you *iemand/iets links laten liggen,* "let it lie on the left."

Similarly, we often use negatively biased language to describe left-handedness, such as *bongo* (Romany, meaning "crooked" or "evil"), *cack-handed* (British English, meaning "excrement-handed"), *canhoto* (Portuguese for "weak," "mischievous"), *gauche* (French for "awkward," "clumsy"), *maladroit* (also French, for "ineffective," "bungling") *gawk-handed* (Scottish English, in which a *gawk* is a foolish person), *kejthandet* (Danish for "cat-handed"), *mancini* (Italian for "crooked"), *molly-dooker* (Australian English, in which a *molly* is an effeminate man and *duke* is slang for "hand"), and *zurdo* (Spanish, in which *azurdas* means "to go in the wrong direction").[18] The English have a particular flair in this regard, referring to left-handers as *back-handed, bang-handed, clickey-handed, coochy-handed, cow-pawed, dollock-handed, gammy-handed, kay-fisted, Kerr-handed, kitty-wesy, scoochy-handed, scrammy, skiffle-handed, skivvy-handed, watty-handed,* and a great number of others.[19] Americans often use the less derogatory term *southpaw,* which according to Leigh W. Rutledge and Richard Donley[20] was coined by Chicago sportswriter Charles Seymour. The typical orientation of older ballparks had the baseball pitcher facing west; therefore, the pitcher's left hand was normally on the south side of the park.

Thankfully, in addition to this long list of derogatory terms for left-handers, there are also some more complimentary exceptions. The Incas called left-handers *iloq'e* (Quechua: *illuq'i*), a word with a positive valence, as the people of the Andes thought left-handers had special spiritual and medicinal capabilities. The Russian term *levsha* for left-hander came from the title character in Nikoli Leskov's 1881 story and has come to refer to a "skilled craftsman."

Although it might seem far-fetched to demonize the left and praise the right in this extreme way, it might come as a surprise to learn that we are still very much influenced by these prejudicial notions during our own everyday tasks. Consider a study by Cornell University psychologist Daniel Casasanto.[21] When asked to choose job candidates from two lists, one on the left, one on the right, the right-handed "bosses" chose more candidates from the list on the right than the identically qualified candidates on the left. Similarly, when asked to sort "good" and "bad" animals into Box A (on the left) and Box B (on the right), the "bad" animals were typically placed into the leftward box (A), whereas the "good" animals went into the right (B).

Even hand gestures during political speeches appear to be vulnerable to this left-right divide. During the 2004 and 2008 U.S. presidential elections, there were two right-handers involved, John Kerry and George W. Bush, and two left-handers, Barack Obama and John McCain. In an analysis of hand gestures during political speeches in the campaigns, the right-handers tended to make right-hand gestures when expressing positive ideas, while left-hand gestures accompanied more negative thoughts. Interestingly, the left-handers demonstrated the opposite pattern. Positive thoughts also followed the dominant (in this case, left) hand, whereas negative thoughts were accompanied by gestures of the non-dominant hand.[22]

Speaking of politics, left and right have very distinct meanings in this sphere, although not so clearly good for one direction and bad for the other, depending on one's own political leanings! The origin of the left-right distinction in politics appears to be derived from the seating arrangements of

France's National Constituent Assembly from the time of the revolution in 1789.[23] The assembly of the time met to decide whether the king should have veto power, and when voting on the matter, supporters of the veto sat on the (noble) right side, whereas those in favour of a more restricted veto sat on the left. This seating arrangement became politically symbolic, with members on the right seeking to preserve the king's power and those on the left wanting to restrict those powers. Throughout the years of the revolution, the left-right distinction was used to describe France's political divide, with the right eventually referring to supporters of an absolute monarchy versus the left signifying backers of a constitutional monarchy. Starting in the 1930s, France's left began to advocate socialism versus calls from the right for economic liberalization. As we will see in subsequent chapters, left-versus-right political affiliations also impact how modern-day aspiring politicians pose for campaign photographs, and differences in posing biases also influence how potential electors perceive political candidates.

TAKEAWAYS

Keeping one's lefts and rights straight can be serious business. If you are lucky, a moment of left-right confusion can result in a relatively harmless navigational error or perhaps a typo in your manuscript. If you are less lucky, an incident of left-right confusion can result in a serious medical error. We use many value-laden terms to refer to our lefts and rights, and in almost all cases, our terminology describing all things leftward tends to be very negative, even derogatory, whereas we have mostly positive terminology for all things rightward. These biases are quite consistent across cultures and time periods. One exception to this general rule can be found in politics. Since the French Revolution, left and right have also had very specific meanings in the political sphere, but the positive or negative valence of right versus left is mostly dependent on an individual's political leanings.

4

Kissing: Are We Doing It Right?

Is this a kissing book?
— *The Princess Bride* (1987)

If you don't remember your first kiss, chances are it hasn't happened yet. Kissing is a big deal. It is rife with ritual and symbolism. Lifelong bonds can be sealed with a kiss and can also be broken with one. If aliens from a faraway galaxy came to Earth to survey the collective artworks of the human race, including our poetry, songs, paintings, and even Instagram posts, they would quickly discover how important kissing is.

However, if they surveyed the collective *scientific* works of humans, they might come to a very different conclusion. Science has not paid much attention to kissing. Ask research psychologists what they can say about what happens when people read the word *green* written in red print and the person is asked to name the colour of the print. Most people are really bad at that, which is a phenomenon called the Stroop effect[1], and research psychologists worth their red ink can report how and why and when that interference occurs and what the effect teaches us about how the brain decodes language. However, ask those same research psychologists about kissing and the answers might well be some blather about hormones such as oxytocin[2] or neurotransmitters such as dopamine,[3] but I seriously doubt anyone would walk away from such a conversation being a better kisser. Interest in kissing might be lost altogether!

Fortunately, the science of kissing is a rapidly growing field.[4] Like every other behaviour surveyed in this book, kissing is typically a lopsided act — for those of us with noses, thank goodness! Unlike many of the behaviours in this book, kissing, normally, requires two people (more on that later), making it even more awkward to talk about than it already is. When kissing a romantic partner, both people tend to turn and tilt their heads to the right (see this effect in Fig. 11, depicting one of the most famous kisses caught on film).

The first major research study to report this bias was conducted by German scholar Onur Güntürkün[5] in 2003, published in *Nature*, which is one of the most prestigious scientific journals in the world. Usually, reading the methods section of a scientific paper is the most boring part, full of details about the technical apparatus employed to collect the data and how the experimental conditions were strictly controlled. Not this time. Güntürkün studied kissing behaviour by observing "kissing couples in public places (international airports, large railway stations, beaches, and parks) in the United States, Germany, and Turkey."[6] Unlike most highly constrained lab-based studies that make it into the pages of *Nature*, this one reads more like the confession of a stalker than a scientific paper. Of the 124 kissing pairs Güntürkün observed, 65 percent turned to the right, compared to only 35 percent twisting to the left. Since this study was published, several other research groups, including my own, have replicated this peculiar finding.[7–12]

In the Introduction, we discussed many of the differences between left- and right-handed people. Do left-handers also kiss right? Absolutely. An interesting replication and extension of Güntürkün's study was performed in Belfast, Northern Ireland, where researchers first observed 125 couples (one

Fig. 11: Rightward kissing biases exemplified on V-J Day in New York City's Times Square on August 14, 1945, captured by U.S. Navy photojournalist Victor Jorgensen (another view of this same scene was originally published in *Life* magazine).

couple more than the German study) kissing each other and then had volunteers kiss a symmetrical dummy/doll.[13] The couples they observed behaved much like the couples in Güntürkün's study, with 80 percent demonstrating a rightward bias when kissing. However, these kisses were coordinated acts between two people. What would happen if it only took one person to kiss? Might the rightward bias be the result of some coordinated action between people? When 240 students were presented with a symmetrical doll and asked to kiss it — hopefully for extra course credit — 77 percent did so turning their heads to the right. Apparently, it doesn't take two to kiss right. These same 240 doll-kissing individuals also took a test of handedness, and there was no difference in handedness between the right and left kissers. So why do we kiss right?

Güntürkün came to a very simple, intuitive but ultimately incorrect conclusion in his study. Correctly noting that humans and other species tend to prefer turning to the right two-thirds of the time, Güntürkün concluded that the rightward bias seen during romantic kissing is caused by this same underlying movement bias. As we'll discover in other chapters, our rightward turning bias appears before birth[14] and therefore cannot be blamed on learning or culture, so it probably influences many of our other lateral preferences, especially those related to how we move around in space, drive vehicles, or even select seats in classrooms, airplanes, or movie theatres. However, we don't *kiss right* simply because we prefer to *turn right*. How come? Because it matters *who* we kiss.

Sharing a mouth-to-mouth kiss with a romantic partner can be a very intimate expression of affection, but kissing a sister doesn't feel the same, nor does kissing a child. It's not that we don't love them, but the subjective feeling of the kiss is very different. As it turns out, the parts of the brain that are preferentially activated during romantic love are a largely different network from that triggered during parental love.[15,16] Parental love stimulates brain structures such as the cingulate gyrus (involved in behavioural regulation) and striatum (involved in movement), whereas romantic love activates structures like the hypothalamus (involved in hormonal regulation) and hippocampus (involved in memory). The huge difference in the subjective feeling of kissing a lover versus kissing one's child should come as no surprise.

Curious about whether the rightward bias seen with romantic kisses would generalize to other types of kisses, my own research group in Canada collected pictures from Instagram, Google Images, and Pinterest of parent-child kisses.[17] Search terms such as *mother kissing son, father kissing daughter* or hashtags such as *#daddykisses* led us to 529 examples of parent-child kisses that met our criteria for inclusion in the study (see Fig. 12 for an example). What happened

Fig. 12: Sample parent-child kiss, not demonstrating the typical rightward bias evident in romantic kisses.

Fig. 13: Romantic kiss demonstrating the typical rightward bias.

to the rightward bias so reliably found for romantic kisses? It was gone, even reversed. We discovered a leftward bias for parent-child kisses regardless of the gender of the parent or child. Might this be because we collected images online instead of observing people at airports or train stations? Apparently not, because when we compared the parent-child kisses to examples of romantic kisses obtained using the same techniques and sources, we found the same two-thirds rightward bias. We concluded that family does matter and that the directionality of kissing biases is modulated by the context of who's kissing whom. If people tended to kiss to the right because of a rightward-turning bias, it wouldn't matter who we were kissing.

Given that most naturally occurring romantic kisses are rightward, what would happen if we reversed that trend? Would an image of a rightward passionate kiss (see Auguste Rodin's *The Kiss*, for example) appear to be less loving or passionate if the figure was mirror-revered? My own research group asked this intriguing question in a few different ways. First, we collected images of rightward and leftward kisses, then presented them along with their mirror-reversed alternatives (see Fig. 14).

Notice that we did not simply take images of rightward kisses and reverse them, because there could be something else different about leftward kisses (more about that later). Instead, we wanted to present people with images of kissing pairs that were originally rightward, reversed rightward, originally leftward, and reversed leftward. Once we had this balanced set of 25 distinct

Fig. 14: An example of mirror-reversed images of rightward and leftward romantic kisses.

image pairs, producing 50 different combinations after we balanced their position on the screen, with half having the original image first, we presented them in pairs to 61 unsuspecting students and asked them to "click on the picture you think displays the most passionate kiss." Can you guess what happened? Just as we predicted, the images that appeared to be rightward kisses were more likely to be selected as the "most passionate kiss." If we think back to that original kissing study conducted in German airports and train stations, the author attributed the effect to the *movements*, the natural rightward turning biases exhibited by most of us. However, that explanation focused on people's biases in *preferred movements* and it does not account for biases in how people *perceive* the passion of rightward kisses differently.

So what did we do next? Just as in our follow-up to the original kissing bias study where we studied parent-child kisses, we also followed up the mirror-reversal study with pictures of parents kissing their children. This time we presented pairs of original and mirror-reversed images to 113 college students and asked them to "click on the picture you think displays the most *loving* kiss." What happened? Just as we found with "naturally occurring" parent-child kisses, the rightward bias we saw for romantic kisses was gone when parents kissed their children.[18]

Another interesting venue for studying kissing biases is through advertising. Many ads — still photos in magazines, billboards, video commercials, online banner ads — depict couples kissing, especially if the advertisement is for a product even loosely connected with romance. I have never seen a commercial for third-party liability insurance featuring a couple kissing, but it is very easy to find smoochy perfume ads. Advertisers work very hard to present products in a way that is compatible with their intended use.

A study by Ryan Elder and Aradhna Krishna of the Marriott School of Management at Brigham Young University[19] examined whether people preferred ads for items that were oriented in a way that made it easy to imagine engaging with the product. For example, they showed a picture of a hand holding a hamburger, measured the preferred hand preference of the person being studied, and assessed whether they favoured images for hamburgers being held in the chosen hand. Over a series of five studies, they found that consumers selected ads for products modelled using their preferred hand.

The follow-up kissing study is obvious. Are ads featuring rightward romantic kisses preferred by consumers and do they impact the consumer's impression of the brand being advertised or even the intent to purchase the advertised item? To study this, we needed to collect "kissing ads" and modify them to produce opposite-kissing variants. This turned out to be a little more complicated than our previous couple of studies, because unlike the Instagram posts of romantic couples kissing, or parent-child kisses, advertisements have something extra in them: words. Of course, simply mirror-reversing the advertising images would not work. The words would be backward on the reversed images. Instead, the text was separated and overlaid to mimic the original layout but without reversing the text and making it obvious which version of the print advertisement was the original. We then presented these ads and their mostly mirror-reversed counterparts and measured the impact on potential consumers. Just as we predicted, ads featuring rightward kisses had higher scores for the attitude toward the ad, view of the brand, and even the consumer's intention to purchase the product.

So far the kisses we have been examining have been between people who know each other well, and the pairs might even be in love. But have you ever kissed a stranger? Now you might be relieved to learn that those of us running experimental psychology studies cannot just do whatever experiment we like. Before starting any new protocol, we need to submit detailed proposals of all the research methods, procedures, goals, and potential participants to ethics boards who work very hard to ensure compliance with ethical, legal, and moral standards. If I were to write up a proposal about asking strangers to kiss each other while I filmed the encounters, my local university ethics board would laugh me out of the room.

Enter the social media phenomenon of "First Kiss" videos on YouTube.[20] Social media trends don't need to conform to the standards set by research ethics boards. If they did, stupid phenomena such as the "Tide Pod challenge" would never infect the internet, nor would they afflict impressionable young minds and bodies. Fortunately, some useful fads emerge online, too. The New York–based clothing company Wren released a short film in 2014 entitled *First Kiss*, directed by Talia Plleva. This movie depicted 20 unacquainted individuals who had consented to be randomly paired with each other in order to engage in a First Kiss video on film. Some, well … many of these pairings are hard to watch. The awkward body language and hesitant initiation of the kiss can be easily inferred by the viewer. Fortunately, some of the pairings come off as natural and even passionate. As is so often the case online, others started copying the original idea, posting their own montages of First Kiss videos on YouTube. This presented us with a unique opportunity. Proposing to have hundreds of people randomly paired to kiss each other would have surely been rejected by any major university ethics board, but the emergence of this social media phenomenon suddenly made the research study possible!

We coded the first kiss encounters of 226 couples for directional biases. Did these couples kiss right? Absolutely not! There was almost a perfect 1:1 ratio between leftward (48.2 percent) and rightward (50.9 percent) kisses, with 0.9 percent of the encounters exhibiting no biases whatsoever (a central kiss). Once again, it matters who is kissing whom. Romantic kisses between loving partners result in rightward kisses. Kisses between parents and children don't exhibit this rightward bias; instead, there might even be a bit of leftward bias. Pair two random adults for a kiss and there's no directional bias whatsoever. Are kissing biases simply due to turning biases? Apparently not.

Other clues about why we kiss right can be found outside the Western world. The text in this book is written in English and is scanned from left to right as the page is read. Most modern languages of Europe, North and South America, India, and Southeast Asia are written from left to right. However, there are some popular languages that read from right to left, including Arabic, Aramaic, Hebrew, Persian/Farsi, and Urdu. For readers who have a native language that reads right to left (RTL), might kissing look different than those from the West who read left to right (LTR)? Remember that the kissing studies I have surveyed so far were from Western cultures such as Germany, the United Kingdom, and Canada.

In 2013, cognitive psychologist Samuel Shaki[21] asked this intriguing question in a couple of different ways, both of which will seem familiar by now. Just as in Onur Güntürkün's study, Shaki observed and coded public kisses by couples, but unlike the original study, Shaki's compared kisses between RTL and LTR reading groups, sampling spontaneous public kisses in Italy, Russia, Canada, Israel, and Palestine. Just as has been observed in other studies, two-thirds (67 percent) of Western couples demonstrated a rightward bias while kissing, but 78 percent of the Middle Eastern couples turned their heads leftward while kissing!

Shaki's study also extended this finding by asking student volunteers to kiss a life-sized symmetrical plastic mannequin's head mounted on a height-adjustable tripod, positioned centrally against a plain background. Students were asked to stand directly in front of the doll's head and kiss the head on the lips. Although the methods section makes no mention of it, I am confident that the researchers had the doll sanitized between trials! Thank goodness for research ethics boards. Just as they observed between "real" kisses with couples in the wild, in the laboratory, the Western (LTR reading) students tended to kiss the doll using a rightward head tilt, whereas the Arabic and Hebrew students were inclined to kiss to the left. Collectively, these results suggest that lateral biases during kissing don't merely depend on the nature of the relationship between the people kissing, but also on the left-right or right-left visual scanning tendencies of the people doing the kissing.

However, there is another type of kissing we have not even considered yet in this chapter: French kissing. No, not *that* type of French kissing. In some countries, like France, cheek kissing is a greeting act, both for hello and goodbye (an aloha of sorts). It happens frequently, and not just between people who love each other, or perhaps even between people who know each other particularly well. Greeting kisses can even be exchanged between people meeting for the first time. However, not every social pairing results in a greeting kiss. They are common between male-female adult pairings, and even reasonably common between female-female adult pairings, but relatively rare between males or children. If we seek social advice on Google or YouTube about how to engage in a greeting kiss in France, we will find advice to kiss the cheek on the receiver's right (our left), and if we are being kissed, we should offer the right cheek automatically. As we will see shortly, the validity of this advice depends on the region of France being visited.

In many cases, these greeting kisses aren't singular acts but rather carefully orchestrated sequences of kisses, alternating between one cheek and the other, in which individuals exchange up to four kisses within a single series. The potential for an interaction to go awry is great. If the local custom is to kiss three times, first right cheek, then left, then back to the right, even getting the initial side wrong can result in thrust/withdrawal awkwardness instead of the "daily miracle of social coordination"[22] that it was meant to be.

As might be imagined, studying the lateral biases during these types of kisses becomes a complicated business. If the local custom is to always kiss one cheek and then the other, which cheek matters more, the first or the last? What if the cultural norm is to kiss three times, starting with the right, switching to the left, and then back to the right? Do we count the right-cheek kiss twice? A very thorough study of this complex phenomenon in France was conducted by Amandine Chapelain and colleagues in 2015.[23] Using a combination of methods, including naturalistic observation (watching and coding real-world, spontaneously occurring kisses in public spaces) and a questionnaire, they found that the majority of regions in France (an LTR-reading nation) demonstrated the same rightward kissing bias reported elsewhere.[24] The first contact in a greeting kiss tended to be rightward, in similar proportions to those encountered elsewhere. However, this practice varied heavily based on region, and within leftward-kissing "departments" (zones in France), people consistently followed the local custom. In other words, the rightward bias noted in most Western societies could be modulated, even reversed, by social pressures and local customs.

Fortunately for the confused traveller, there are some relatively steady regional trends regarding the laterality of greeting-kissing biases in France. There is a website called *combiendebises* ("how many kisses" in English) where travellers can research how many greeting kisses are normal for a particular region and which side should be kissed first (combiendebises.com).

At this point in our kissing journey, we have answered a number of questions but also posed new ones along the way. What causes the rightward kissing bias among Western couples? It isn't simply a turning bias, because it matters who is kissing whom. If it is a romantic kiss, it will probably be rightward; if it is a familial one, probably not. Is the rightward bias related to hand preference? Apparently not. Is it caused by native reading direction? Possibly, at least it appears as though one's native reading direction can

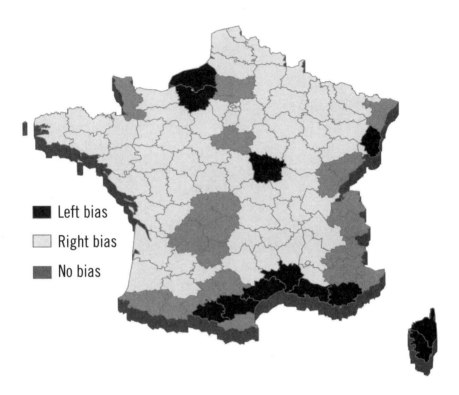

Fig. 15: Kissing-bias observations recorded in France. Note that the majority of the country exhibits rightward biases (white regions) and that the minority of regions where leftward kissing is the norm (dark regions) can be found in the south of France.

influence the effect. Similarly, the local social customs and social pressures, at least within different regions of France, can also modulate the laterality of greeting kisses.

When my research group's study about familial kissing came out, much of the media coverage included the same amusing and intuitive mistake. Outlets such as *Cosmopolitan* or *Maxim* were eager to use the rightward bias effect as a diagnostic for how romantic a kiss is between a couple, and even as an indicator of the trajectory or fate of one's relationship. If one's mate turns to the right when kissing, all is well. However, when facing a leftward kiss, the friend zone has been entered! At best those claims are as yet unfounded, and at worst, they're just plain wrong. I can see the logic behind those predictions and can even imagine ways of testing them. However, until that experimental work is done, those leaps are premature.

TAKEAWAYS

International Kissing Day takes place on July 6 every year. How can we prepare for this momentous day? What practical advice can I offer about kissing? We know that rightward kisses are perceived to be more romantic than leftward ones. When kissing a lover, turn to the right. The only possible exception to that suggestion would be if both people in a couple are from cultural backgrounds with a first language that reads right to left. When kissing a family member, central or even leftward kisses are the norm. When visiting France, or another part of the world that enthusiastically adopts greeting kissing, look up the pattern of greeting kisses appropriate for the region before attempting any, or at the very least, let the other partner make the first move before thrusting or jabbing inappropriately. As the Swedish actress Ingrid Bergman once said, "A kiss is a lovely trick, designed by nature, to stop words when speech becomes unnecessary."

5

Cradling Biases: Are You Holding Your Baby Right?

Tell me where is fancy bred,
Or in the heart, or in the head?
WILLIAM SHAKESPEARE, *The Merchant of Venice*,
act 3, scene 2

O f all the lopsided behaviours we are exploring, cradling biases are probably the oldest. Our tendency to cradle infants to the left is *old* in several interesting ways. If we compare our own lopsided behaviour to that of other animals, humans are unique in many of our lateral biases. It should come as no surprise that the human biases present in our artwork, seat selection, gestures, politics, and social media are nowhere to be found elsewhere in the animal kingdom. But what about some of our more simple and pronounced biases, like handedness? Sure, a cat or dog might exhibit a preference for one paw or the other when offering to shake a paw (dog) or pushing that heirloom vase from grandmother off its shelf (cat), but the 90 percent right and 10 percent left population-level split exhibited by humans is nowhere to be found elsewhere in the animal kingdom. Instead, individual cats and dogs display relatively weak lateral biases, and at the level of the species (or even breed), the biases are even weaker still.

Fig. 16: Stone statue of the Virgin Mary cradling Jesus to her left, the same bias exemplified by most people today.

Not so with cradling. Take a field trip to the local zoo and observe clear leftward cradling biases in monkeys, chimpanzees, and a host of other species. Observing the same behaviour across species is usually of great interest to scientists because it suggests something *adaptive* about the behaviour; something special about that behaviour enables the animal to survive and reproduce. In the case of cradling, the same leftward bias displayed by humans (see Fig. 16 for a sample of the leftward cradling bias) can be found in animals that roamed the planet hundreds of thousands of years before the earliest humans appeared and started wrecking the place. Cradling biases are old.

Research into cradling biases is also ancient. By some accounts,[1] Plato was the first to document cradling biases when discussing handedness in *Laws*. We can interpret the Greek philosopher's claim that the future handedness of children could be blamed on the "folly of nurses and mothers" as a reference to cradling biases (as the "folly" in question) in part because of another passage in *Laws*.[2,3] Here, Plato suggests that nurses should carry infants "to the temples, or into the country, or to their relations' houses," but they should "take care that their limbs are not distorted by leaning on them when they are too young".[4] Personally, I am not convinced that Plato was the first to describe cradling biases.

Other early references to cradling biases are more explicit. For example, *The Children's Book* (1656) by Felix Würtz, a Dutch physician and surgeon, suggests that always carrying the child on the same side could be "hurtfull also unto children."[5] Another European surgeon also "faulted" nurses and mothers with the blame for an infant's handedness based on cradling biases. Nicolas Andry's 1741 *L'Orthopédie* claims that left handedness is "commonly owing to a fault in Nurses, some of whom carry the Children always upon the left Arm, by which means the [child's] left Arm is only at liberty, and so they employ it upon all Occasions; whence the left Hand becomes stronger, and the other weaker."[6] Other scholars in the 18th and 19th centuries, such as philosopher Jean-Jacques Rousseau[7] (1762) and Joseph Comte (1828), noted cradling biases and pontificated about the potential influence on the development of handedness in infants. However, after child psychologist Lee Salk visited the Central Park Zoo in New York City 50 years ago, he triggered decades of research into cradling biases.

When Salk watched a rhesus monkey cradle her newborn, he noticed her "marked tendency" to cradle the infant to her left side. Over the coming weeks, Salk recorded further cradling biases, counting another 39 leftward holds and only two on the right (a 95 percent leftward bias). Following up on these curious observations, he wondered if human mothers exhibited the same bias. Instead of merely studying human mothers "in the wild," Salk devised a series of experiments he could carry out in the maternity ward of his local hospital during the first four days after birth.[8] He picked up an infant from behind using both hands and presented it centrally to the mother. Then he recorded how she chose to initially hold the child, finding that right-handed mothers cradled to the left 83 percent of the time and that left-handed mothers also cradled to the left, although the bias was a little weaker (78 percent of the holds were leftward).

When Salk asked the mothers why they cradled to the left, the answers varied according to the handedness of the mother. Left-handed mothers told Salk, "I'm left-handed and can hold my baby better this way," whereas right-handed mothers claimed that "I'm right-handed, and when I hold my baby on the left side it frees my right hand to do other things."[9] Instead of concluding that the mothers were doing the same thing for different reasons, Salk thought these explanations were simply rationalizations for an automatic response that had nothing to do with handedness. The research that ensued mostly supports this view. Handedness of the mother appears to have little if anything to do with the cradling biases we see in humans or even in other species, but we will get back to that later in this chapter.

After this surprising study, Salk began to pay special attention to paintings and sculptures of mothers with their children. He observed 466 such works of art, including Renaissance-era paintings of the Madonna and Child or three-dimensional sculptures of mothers with their children. Regardless of the medium of the art or the subject matter, 80 percent of the pieces depicted leftward cradling,[10] very similar to the biases observed by modern mothers in the real world. This same leftward cradling bias is also obvious in early Christian art and Impressionist and Post-Impressionist paintings,[11] although the bias tends to be weaker or completely absent in pictures of men holding babies. Thanks to a study of pre-Columbian

American art,[12] we know that works from as early as 300 BCE exhibit a leftward cradling bias (see Fig. 17).

The obvious question then is why? Of course, Salk himself asked this in his first study, and the answers he received didn't help much. Since Salk's original report, more than 50 follow-up studies have found the same leftward cradling bias.[13] We can see it in maternity wards, public parks, works

Fig. 17: Example of a Mayan female figure holding her child to her left.

of art, private homes, even on Instagram. But these new studies have also sparked new questions. The bias is strongest with very young infants, and by the time the child reaches three or four years old, the bias can disappear completely or even reverse. Both men and women show the bias, but it is stronger in the latter. The leftward bias can be found almost anywhere in the world, including the Americas, Europe, Africa, and China, but there are some corners of the globe where it disappears, as exhibited by the Malagasy people of Madagascar.[14] Any successful explanation of the cause of the bias would also have to explain variations in the bias.

So let us start with the most obvious explanation and the reason for the William Shakespeare epigraph that opens this chapter. When one of Portia's attendants sings, "Tell me where is fancy bred, / Or in the heart, or in the head?" in *The Merchant of Venice*, she is referencing what modern psychologists now call the cardiocentric hypothesis, the idea that intellect and emotion come from the heart, not the brain. Aristotle is widely credited — or blamed? — for this position, but he was not the first or the last to credit the heart with higher functions unrelated to pumping blood.[15] He noticed that bodies became cool after a person died and ascertained that the heart must be the source of the heat, even crediting the brain with the function of cooling off the body. Many of Aristotle's contemporaries, such as Empedocles, championed this cardiocentric view, whereas Plato, Democritus, and eventually Galen argued that the brain is, in fact, where "fancy is bred," playing a central role in intelligence and emotion.[16]

Despite the cardiocentric hypothesis being dead for centuries now, its influence lives on in, and arguably dominates, our everyday language and symbols, especially when we talk about emotion. We tell a lover "I love you with all my heart," we give them "heartfelt thanks" for kind deeds and gifts, and we mourn our "broken heart" if the lover leaves us. A person who appears to behave devoid of emotion has "a heart of stone," and if we say something really profound and personally meaningful, we "speak from the heart." The heart does not just get credit for emotion. When we memorize a song or a written passage from a famous Shakespeare play, we "learn it by heart." Our symbols of love and affection also give the heart all the credit. When is the last time we saw a Valentine's Day card with a picture of a brain on it?

Which brings us back to Lee Salk working in New York City in the late 1950s. Salk, too, thought it odd that science was happily giving parts of the brain, especially the hypothalamus at the time, all the credit for emotion, while our everyday expressions were focused on the heart. Specifically, he wondered if the saying "close to a mother's heart" was more than just an expression and was perhaps the basis for behaviours such as the cradling bias seen in humans and monkeys alike.

Except for some very rare examples of situs inversus[17] (see Fig. 18), the human heart is normally positioned to the left. Cradling infants to the left puts them literally "close to the heart" of mothers. In utero, in the womb, the unborn child hears the sound of the mother's heart throughout development. Salk argued that children would normally learn to associate this sound with a secure and relatively stress-free environment. Therefore, placing the relatively newborn child next to the heart could presumably help

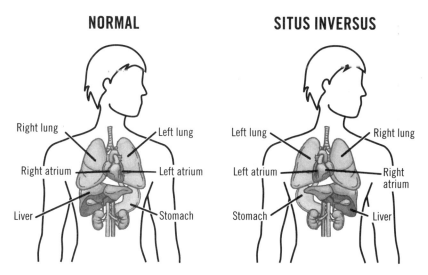

Fig. 18: The rare (1/10,000) condition of situs inversus totalis in which the normal asymmetry of the internal organs is reversed, including a rightward displacement of the heart. Most people with situs inversus go through life without any complications from this unusual condition.

pacify the baby, making him or her feel more secure. Among the various versions of this argument, Salk even hinted at a bolder and broader version: "from the most primitive tribal drumbeats to the symphonies of Mozart and Beethoven there is a similarity to the rhythm of the human heart."[18]

As intuitive as the heartbeat hypothesis is, the evidence for it is weak. Mothers with dextrocardia — the heart positioned to the right instead of the left — still cradle to the left.[19] Even mothers with "their hearts in the right place" don't tend to cradle babies in the pericardial region where the heartbeat is most clearly audible. Salk's original studies on cradling biases included some very striking claims about how soothing the sound of a heartbeat is to an infant, but attempts to replicate those results have failed.[20]

There is an even simpler way to test the heartbeat theory. We can look at how people carry things that aren't babies. For example, how about shoppers lugging packages that are somewhat baby-ish in size and shape? At the University of Southern California, I. Hyman Weiland observed shoppers as they transported "baby-sized" packages through doorways with automatic doors (not requiring a free hand to open the door). Of the 438 adults he observed, exactly half held the package on the left, the other half to the right.[21] Similarly, if hundreds of college students imagined holding a vase, a book, the same book in a paper bag, or a sealed box filled with powdered sugar, there was no side bias whatsoever.[22]

However, a real baby is not necessary to trigger a leftward cradling bias; a doll will do. In fact, even a *real* doll is not needed; an imaginary one is sufficient. If we ask college students to imagine holding a baby or doll, the students tend to visualize grasping the doll to the left.[23] Similarly, if we ask a number of adult women to clutch pillows against their chests, we shouldn't see a holding bias. However, if we request that same group of women to picture the pillow as an "endangered infant," the leftward bias returns.[24]

The term *furbaby* (and *#furbaby*) is growing in popularity, along with the companion term *pet parent*, but *furbaby* is not just another synonym for *pet*. Instead, the parent of a furbaby typically doesn't have any human children, and the pet is treated in many ways as a stand in for a child. If pretending that a pillow is a human baby is adequate to trigger a left-sided cradling bias, how about furbabies? Are they, too, treated like "real" babies?

Fig. 19: Supermodel Miranda Kerr cradling her furbaby to her left at Los Angeles's airport.

Yes, indeed, they are! In a survey of both celebrity and "regular" dog owners, 62 percent of the women pictured cradled the dog to the left.[25] In my own lab, we surveyed more than 1,000 images of parents cradling their children and regular, everyday (non-celebrity) folks holding their pets and found that just like the celebrities, normal folk also prefer to cradle furbabies to the left.

So far we have discussed cradling biases as if they are the same for everyone and are perfectly consistent over time. Neither of those things is true. There are people who consistently cradle infants to the right (or at least fail to show the typical leftward bias), and even with people who do cradle to the left, the bias typically changes over time, depending on the age of the infant. The bias is strongest when cradling newborn and very young infants,[26] but as the child ages and becomes larger and heavier, the bias declines or even reverses.[27] Interpreting this age effect is difficult, since there are several important differences between a newborn and a three- or four-year-old that could influence how the child is held. Newborns are small, light, fragile, and not very strong; they even need neck support. By contrast, three- or four-year-old children can weigh four to six times as much as a newborn, and parents might switch sides more regularly just to combat fatigue.

Other factors also influence the strength and direction of the holding bias. Mothers who have been separated from their infants due to illness or a premature birth demonstrate weaker leftward cradling biases than mothers experiencing uncomplicated births (if there is such a thing as an uncomplicated birth).[28] There are also some interesting differences in the qualitative experience of left versus right cradlers. New mothers who cradle to the left have higher "feelings of affinity" with their infants.[29] Left-cradling mothers also report making more preparations for the child prior to delivery.[30]

This would imply that rightward cradlers have a tougher time, and there is plenty of evidence for that, too. For example, maternal depression is related to rightward cradling.[31] The Beck Depression Inventory (BDI)[32] is a short, 21-item questionnaire that asks about attitudes and symptoms of depression, including mood and changes in sleep or eating habits. Psychologist Robin Weatherill[33] gave the BDI to a group of 177 high-risk mothers, half of whom had experienced domestic violence with their partners. The non-depressed mothers demonstrated a strong leftward cradling bias, but this bias disappeared in the depressed mothers and even shifted to a slightly rightward one. Of course, that raises the "chicken-and-egg" question. A study by Peter de Château and colleagues showed that mothers who cradle to the left have a higher "feeling of affinity" with their infants.[34] This feeling could possibly lead to depression or maybe vice versa.

The lack of leftward cradling bias in depressed women also fits nicely with what we already know about the etiology of depression. Depressed people generally show right-hemisphere dysfunction, including less activation of the hemisphere overall,[35] less perceptual reaction to visual stimuli presented to the right hemisphere, less reaction to emotional content presented to the right hemisphere, and decreased response to positive emotional stimuli presented to the right hemisphere. All of this is consistent with a lack of leftward cradling biases.

Cradling biases also appear to be dissimilar in people with autism spectrum disorder (ASD). Children with ASD often exhibit difficulties bonding to others and adopt a different approach to emotional relatedness compared to neurotypical children. A recent study employed a "pretend-play" task with a group of 20 children with ASD compared to 20 neurotypical children.[36] All children were handed a doll (named Suzie) and asked, "Will you hold Suzie like you are putting her to sleep?" and the side of the hold was recorded. Ninety percent of the neurotypical children demonstrated the expected leftward bias, but the kids with ASD displayed no bias whatsoever, with 50 percent leftward and 50 percent rightward holds.

These leftward cradling biases appear regardless of handedness, culture, or ethnicity, but what about scenarios when the baby and child have unalike ethnicities? In a clever and somewhat disturbing study, a group of Italian researchers had Caucasian women cradle white or Black dolls, then assessed their level of prejudice toward African individuals.[37] The more prejudiced the cradlers were against Africans, the more the cradling preference shifted away from the leftward norm. The less ethnic prejudice exhibited by the cradlers, the more likely they were to have a leftward cradling bias. Collectively, these findings suggest that leftward cradling is a natural index of attachment and positive relationships between an infant and parent.

So we know that most people cradle real, imagined, and even "fur" babies to the left, but how about other animals? Do they also exhibit this lopsided care of their offspring? Remember that Lee Salk's popular study from 50 years ago started with his observation of a macaque monkey in the Central Park Zoo in New York City that cradled her infant to the left 95 percent of the time. Since that observation of a single monkey in captivity, many follow-up studies of different species in the wild and in captivity have confirmed Salk's observation.

Chimpanzees seem to have the strongest leftward bias, averaging near 75 percent.[38] Gorillas also cradle their infants to the left much of the time (74 percent), but the bias is weaker or gone entirely in gibbons, orangutans, and baboons. Studies of monkeys have produced a mixed bag of results.

Fig. 20: Chimpanzee mothers cradle to the left 75 percent of the time, almost as frequently as humans do.

Of course, Salk reported a really strong leftward bias from a single captive macaque monkey in the zoo, and some follow-up investigations have discovered leftward biases for infant carrying[39] or in picking up infants when the mother is frightened.[40] Others have found biases at the level of the individual but not the entire group of monkeys being observed.[41] However, none of these studies report biases as strong as the ones in Salk's report. Overall, though, it appears as though the great apes, including us, cradle to the left.

It isn't just the cute, cuddly, furry animals that cradle to the left. Even animals such as fruit bats (see Fig. 21) have strong leftward biases for mother-offspring pairings. Infant bats spend much more time attached to the left nipple, although not necessarily suckling all that time. Just as is the case with human pairings, this arrangement has both animals keeping the other in the left field of view, preferentially exposing the right hemisphere.

How about other animals? Do we see the same leftward biases between mother and child in other species? What about animals that don't cradle at all? Consider the unusual example of feral horses. A couple of recent studies describe strongly lateralized mother-offspring behaviours in domesticated

Fig. 21: An infant Indian flying fox fruit bat attached to its mother's left nipple.

horses. When mares (mothers) and foals (offspring) reunite after a separation, the foals much prefer positions that keep the mother in their left fields of view. Similarly, the mares also appear to favour keeping their foals to their left sides, but only when fleeing from some perceived danger.[42] This kind of positioning keeps the foal in the part of the mother's visual field dominated by her brain's right hemisphere. We know that the right hemisphere is crucial for various social behaviours in horses, including bonding. The right hemisphere also plays a crucial role in spatial processing, perceiving location, the location of objects, and the relative position between the viewer and those objects. The right hemisphere seems to be ideally suited to track offspring!

Not every species cradles to the left, though. Most primates do and so do some bats, horses, the southern right whale, and maybe even reindeer. Despite the perfectly posed leftward cradle at the population level, the walrus does not exhibit a cradling bias. Neither do antelopes, beluga whales, eastern grey kangaroos, or muskox. Some animals even appear to cradle rightward, including argali and red kangaroos.[43-46]

At this point in our survey of cradling biases, I hope I have convinced the reader of a cross-cultural and maybe even cross-species leftward cradling bias. Even more astonishing, it doesn't seem to be learned. If we hand a newborn baby to a teenage boy who's never cradled a child in his life, chances are, he will cradle to the left. The big mystery is why. Lee Salk's original theory about keeping the child close to the mother's heart is appealing in its simplicity, but the evidence accumulated in the 50-plus studies since his fateful trip to the zoo hasn't provided much support for that idea.

Another possibility is that leftward holds create a greater sense of intimacy and affinity between the baby and parent. Putting the infant in the parent's left side of space and left visual field could allow for more engagement of the right hemisphere. It also exposes the child to the left side of the parent's face, and as we will learn in Chapter 6, leftward poses are more emotional than rightward ones. While this explanation is a bit clumsier than the "mother's-heartbeat" theory, it does a better job of accounting for many, but not all, of the studies in this area. It certainly does a better job of explaining why maternal horses keep their foals to their left, especially when threatened. It also helps explain why a depressed mother might be less likely to cradle to the left, or why someone on the autism spectrum who might be less inclined to seek physical intimacy and affinity with a baby is also less likely to cradle to the left.

TAKEAWAYS

Overall, this pattern shows us that the human tendency to cradle infants to the left is not uniquely human at all. It is widely seen throughout other primates and other mammals, perhaps because of the perceptual and emotional pair-bonding benefits provided by the arrangement. How should we cradle babies? The right way is usually left.

6

Posing Biases: Putting the Best Cheek Forward

Who sees the human face correctly: the photographer, the mirror, or the painter?

— PABLO PICASSO

You probably know the old joke that "when you start to look like your passport photo, it is time to go home."[1] One can find plenty of variants on that joke, too, with endings such as "you probably need the trip" or "you are too sick to travel," but they all imply the same thing: Your passport photo probably looks terrible. Passport photos are almost universally awful. Don't smile, don't wear a hat, don't wear glasses, don't don any other facial obstructions, stare straight into the camera, and *click!* A photo we have to carry with us for years. Going to the Ministry of Transportation or posing for an employee photo ID yields pretty similar results. Portraits of people looking straight at the camera are not very flattering.

If we scroll through the profile pictures of friends on Facebook or look at potential suitors through a favourite dating app, and we will notice that people rarely pose for photographs straight-on. We almost always turn to one side or the other. Celebrities accustomed to having pictures taken of them often exaggerate this tendency for lateral posing. Just watch red-carpet footage on YouTube from a recent award show; the lateralized facial posing will be almost as obvious as the fake tans and fake smiles.

Like every other example surveyed in this tour through lopsided human behaviours, posing is not an equal-opportunity act. If posing direction was random, 50 percent of our non-central selfies or profile pictures would predominately feature our right cheek, with the other 50 percent displaying the left. But they don't. Most pictures of human faces portray a left-cheek bias, including the most famous portrait of all, the *Mona Lisa* (see Fig. 22). These images can be hundreds of years old and hand-painted by one of the great masters, or they can be the most recent, haphazardly Snapchatted selfie from a teenager at a mall. It doesn't matter if the image is crafted by hand, captured with a professional camera, or taken with a phone. The left-cheek bias can even be found on coins. When the British Royal Mint presented designs to King Edward VIII (the Duke of Windsor at the time), he rejected the ones that showed the right side of his face because of his perception that the features on his left side were "superior."[2] Only under very special circumstances will we ever find a right-cheek bias in a group of photographs, and I'll describe those conditions later in this chapter. So what's going on? Why do we turn the left cheek for a portrait?

Charles Darwin was among the first to describe leftward biases in how we express emotion, especially the more aggressive expressions such as sneering, typically revealing the canine tooth on the left side. In 1872, he published *The Expression of the Emotions in Man and Animals*,[3] a work obviously overshadowed by his more famous book, *On the Origin of Species*.[4] However, his work on emotion also sparked countless scientific studies. One of the more famous examples is a series of studies by American psychologist Paul Ekman.[5] Before Ekman's work, scientists widely believed that humans learned how to communicate with one another through gesture, speech, and expression, and that different cultures demonstrated huge variation in

Fig. 22: Leonardo da Vinci's *Mona Lisa* demonstrates the left-cheek posing bias in portraiture.

their languages and communication styles, including facial expressions.[6] Although it is certainly true that we learn language and culture, there also appear to be behaviours we are born with.

Studying an isolated tribe in Papua New Guinea, Ekman discovered that even a geographically and culturally isolated people exhibited the very same facial expressions as all other humans on Earth, and he described the universal emotions and their accompanying facial expressions, including happiness, surprise, sadness, anger, disgust, contempt, and fear (see Fig. 23).

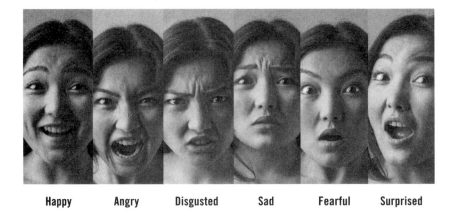

| Happy | Angry | Disgusted | Sad | Fearful | Surprised |

Fig. 23: The six universal facial expressions described by anthropologist Paul Ekman.

These basic emotional expressions have been studied extensively, and many different research groups have conducted numerous lab-based experiments that found the left side of the face is more emotionally expressive than the right.[7-11] This is because the right hemisphere of the brain dominates emotional processing and also has more control over the left side of the face (the lower two-thirds).[12]

At Cambridge University in the United Kingdom, Chris McManus (one of my own academic mentors and former supervisors) and Nicholas Humphrey[13] were the first to report that famous portraits demonstrate a left-cheek bias. They studied 1,473 formal portraits done between the 16th and 20th centuries available at the National Portrait Gallery in London, the Fitzwilliam Museum in Cambridge, and several anthologies of portraits (see Fig. 24 for examples). Most of the painted subjects were posed with a rightward turn (showing the left cheek), an effect that was more prominent in representations of females (68 percent left cheek) compared to those of males (56 percent left cheek). Keep an eye on that difference between males and females because it comes up again soon, then also later in this chapter.

Fig. 24: Portraits of James I and Elizabeth I. James puts the right cheek forward, whereas Elizabeth turns the left cheek.

Why do painted portraits tend to highlight the left cheek? First, let's review a "mechanical" explanation that focuses on the artist instead of the subject. Consider the scene depicted in Johannes Vermeer's *The Art of Painting* (see Fig. 25). Right-handed painters might normally prefer to place subjects to their left so that the subjects are less likely to be blocked by the easel or canvas. Artists as such would present their left cheeks to their subjects, who in turn would reciprocate.

Another possibility is that the handedness of the artist is to blame. Despite the fact that many famous artists from the recent past were left-handed, the vast majority were right-handed, and it might be easier for a right-hander to paint a leftward-facing profile, just as it is easier for right-handers to write words left to right.

If these explanations are not convincing, that's not surprising. Neither explanation accounts for gender difference in the leftward bias, in which females are more likely to display the left cheek. Furthermore, if the handedness of the artist is the cause of the posing bias, then a reversal of the handedness of the artist should result in a reversal of the bias. Many famous painters

Fig. 25: Johannes Vermeer's *The Art of Painting* depicting how right-handed painters typically position the subject of the portrait to the left.

were left-handed, including Leonardo da Vinci, Rembrandt, Michelangelo, Raphael, Hans Holbein, M.C. Escher, Vincent van Gogh, and Peter Paul Rubens,[14] but a survey of the side biases exhibited by their portraits reveals that same left-cheek bias evident in right-handers. For example, 70 percent of Raphael's portraits present the left cheek, as do 57 percent of Holbein's.

Therefore, hand preferences of artists are unlikely to be to blame for the left-cheek bias.

Instead of counting cheeks in old paintings, there's another way to rule out these "mechanical" explanations. We can study portraits taken with a camera. Here, too, regardless of whether the picture is snapped with a professional-quality camera (one typically held with both hands) or a cell-phone (usually grasped with one hand), the leftward bias is present. Clearly, in those cases, it has nothing to do with drawing curves or the position of the easel. Furthermore, the stronger left-cheek bias for portraits of females is *even stronger* for portraits taken with cameras compared to paintings, suggesting that some other factor is at play. But what? If the left bias isn't an *output* effect (i.e., a mechanical bias), might it be an *input* effect?

Maybe people prefer to *look* at portraits featuring the left cheek. In other words, perhaps it is a *perceptual* effect. Within the right hemisphere of the brain, there are regions, such as the right fusiform face area (rFFA), that appear to be specifically dedicated to processing images that look like faces, including the identity of and emotional expression in the face.[15] When we think we see a face in the clouds, in a burnt grilled cheese sandwich, or in the foam of a latte, that is also the result of brain activation within the rFFA. This region of the brain appears to go through life asking, "Where is the face? Where is the face? Where is the face?" all day.[16]

In the lab, it is easy to demonstrate that we pay more attention to the left cheek (which falls in the right field of view) when perceiving faces. See Fig. 26 for an example of a chimeric face, which is a composite image made up of two different pictures split down the middle. The image doesn't even have to be the same person. A chimeric face can be from two individuals, or even two different facial expressions from the same individual. If these weird-looking face composites are flashed briefly on a screen, people typically identify the one presented to the right hemisphere of the brain, the face on the left side of space. Left-cheek poses expose more of the facial features to the observer's right hemisphere, which makes the face easier to recognize.[17] Left-cheek poses can even be identified more quickly than right-cheek ones.[18]

There are several interesting exceptions to the left-cheek rule. One of them is the self-portrait. If we survey art collections in museums and select

Fig. 26: An example of a chimeric face. If this image is flashed very briefly while viewers look at the middle of the image, they'll typically identify/remember the right side of the image (portraying the left cheek).

only the portraits of artists painting *themselves*, they tend to show the right cheek instead (see Fig. 27). This reversal in the normal bias can be found in portraits painted from the 15th to the 19th centuries,[19,20] but it seems to disappear in the 20th century, when modern photography entered the equation. So what was going on between the 15th and 19th centuries? How were these self-portraits being painted, and why did they feature a reversal of the normal leftward bias?

Mirrors, and the image reversal they cause, are almost certainly part of the reason. If the artist poses in front of a mirror and adopts the "normal"

Fig. 27: A self-portrait of Pablo Picasso at 15 years old, showing the right-cheek bias instead of the left-cheek bias normally observed in portraits of others. Is this reversal because the portrait was painted in a mirror?

left-cheek pose, that mirror image would put the more forward left cheek in the right side of space and the reversed image is what gets reproduced in the self-portrait. So, in a kind of backward manner, the right-cheek effect we see in some of the earlier painted self-portraits might actually reflect a left-cheek posing bias into the mirror. Also, consider the mechanics of creating a 15th-century self-portrait without a photograph of oneself. The self-painting right-handed artist could be placing the mirror to the left so that it is clearly visible, but out of the way of the painting hand, and the artist is not occluding the mirror while raising the hand to paint.

There are some even more complicated potential explanations for the reversal. Because emotional expression in the face is dominated by the right hemisphere of the brain, the left side of the face is likely to be more emotionally expressive. The self-portrait artist might be trying to paint the more expressive side of the face, the left, which appears as the right side in the mirror!

Self-portraits used to be rare. Today, the average teenager can take a dozen of them in a single day, but we don't call them self-portraits. We call them "selfies." Studies of selfies on Instagram and elsewhere indicate a left-cheek bias,[21] but the effect appears to depend on how the picture is taken. In a major survey of 3,200 selfies taken in five major cities (New York City, São Paulo, Berlin, Moscow, and Bangkok),[22] an Italian research team selected 640 selfies from each place and looked for the left- or right-cheek effect but found neither. When the Italians counted all 3,200 selfies, there was no bias toward either side. However, the *style* of the selfie had a huge impact on which cheek was showing. When selfies were taken in the mirror, they tended to feature the right cheek (70 percent of the time). The "standard" (non-mirror) selfies had the opposite outcome; they featured the left cheek but only barely (53 percent of the time). The effect was quite consistent across gender and across the five cities, with small changes to the pattern only evident in Bangkok females and Berlin males.

What is going on? There seems to be a culture-independent preference for displaying the left cheek. Mirrors definitely confuse matters, but they don't reverse the bias. They just reverse the expression of the bias. The database of 3,200 selfies used by the researchers is publicly available at selfiecity.net for anyone who wants to test their own predictions.

The typical left-cheek bias disappears under some other special circumstances. When posing for a formal portrait, academics don't demonstrate the typical left-cheek bias. In a study of formal portraits of members of Britain's Royal Society, the left-cheek bias disappeared,[23] but the same research group found it in other portrait collections, such as those in the National Portrait Gallery. That could be due to the desire to appear unemotional and rational that motivates scientists to adopt a more rightward pose. This strategy appears to work. In a study of how people perceive the

Fig. 28: Examples of professors from the University of Utrecht used in Carel ten Cate's 2002 study. On the left is a right-cheek portrait of Professor Houck, which received a high "scientific" score. On the right is a left-cheek portrait of Professor Wesseling, which received a lower "scientific" score.

portraits of scientists, Carel ten Cate, a professor of animal behaviour at Leiden University in the Netherlands, had people rate painted portraits from 1710 to 1760 according to how "scientific" the professor was.[24] Sure enough, the professors painted with a left-cheek bias were perceived as less scientific than those with a right-cheek preference. In a clever twist, ten Cate also presented mirror-reversed portraits to make sure there was nothing else visually different between the left- and right-cheek pictures, but that made no difference to the subjects in the study. Right-cheek portraits were more scientific regardless of whether the portrait was mirror-reversed or not.

As it turns out, you don't even have to be a *real* scientist to pose like one. In a clever twist on the previous studies of academic portraits, an Australian research group[25] had people pose for a photograph. They were told "You are a successful scientist at the pinnacle of your career … you have just been accepted as a member of the Royal Society and have been asked to provide a portrait for their gallery … you want to give the impression of an intelligent, clear-thinking person … try very hard to avoid depicting any emotion at all."[26] The people in this group given such instructions tended to turn their right cheeks for the portrait, opposite the normal left-cheek effect. Is this reversal because of the artificial circumstance perhaps? Apparently not. In the same study, another group was given a more *emotional* condition, and told "You have a close-knit family … you are going overseas for a year and want to have a portrait taken as a gift … put as much real emotion and passion into the portrait as you can."[27] Participants in this condition presented the left cheek, just as we would expect for an emotional portrait.

Researchers in Japan recently added a clever twist to this study.[28] They wondered whether posers were *aware* of their tendency to pose one way or another. Using the same two conditions (expressing emotion for a family portrait, or showing a calm and reassuring attitude as a scientist), after replicating the original result, the researchers measured whether people were consciously aware of their own biases, or if it was an intuitive, unconscious habit that we produce without awareness. Interestingly, the student volunteers in the study were completely unaware of their own biases or how the type of portrait they were posing for influenced their choices.

Collectively, these studies tell us that when people want to convey emotion in a picture, they tend to present their left cheek, which is more emotionally expressive and dominated by the right hemisphere of the brain (the more emotional half). When people want to hide their emotions or appear impassive, they offer the right side of the face, controlled by the less emotional left hemisphere of the brain.

This posing strategy plays out in lots of interesting ways in the real world. If we compare academics of different disciplines, professors of English are more likely to exhibit left-cheek poses than are scientists.[29] If we contrast male and female doctors, female ones are more likely to show the left cheek than males.[30] Even if we show people a picture of a student and ask them to guess the academic major (given the options of chemistry, English, or psychology), the students displaying the right cheek are more likely to be identified as chemistry majors, whereas those presenting the left check get identified as English students.[31]

So far we have been pretending that people behave consistently across situations, but of course that isn't true at all. Human behaviour always depends on the situation, and posing biases are no different. Perhaps some of the strongest and most common portrayals of cruelty and pain are images of the Crucifixion of Jesus. Christianity places a lot of emphasis on the disgrace and agony of Jesus's death, and there are multiple accounts of the public mocking and taunting that accompanied the scene of his death. How might this emotionally charged scene be presented by artists?

The leftward bias normally seen in portraits is greatly exaggerated in images of Jesus on the cross. In a recent study of pictures that met

Fig. 29: The crucifixion of Jesus. Ninety percent of images of Jesus on the cross show the left cheek.

inclusion criteria (face-forward images in paintings, not reliefs or other media), 90 percent of the depictions of the Crucifixion featured the left cheek of Jesus,[32] a bias far higher than those found in comparable portraits from the same time period. Perhaps artists exaggerated the normal leftward bias in an effort to amplify the emotional expressiveness of this extreme moment.

In addition to these brain-based explanations, we can also consider biblical ones. Many images of Jesus on the cross also feature the Virgin Mary to the right of Jesus at the foot of the cross. Perhaps the rightward turn and exposed left cheek are related to Jesus turning toward Mary. Another biblical explanation relates to other portraits of Jesus and Mary together. In images portraying Mary carrying baby Jesus, she is typically cradling him in her left arm, with his left cheek facing outward in the picture. Perhaps these images of the Crucifixion are simply being consistent with paintings from earlier in the life of Jesus.

Probably not. No survey of artworks depicting the life of Jesus would be complete without considering portraits of Jesus *after* the Crucifixion, depicting the resurrected Jesus. In a subsequent study, the same American group that performed the Crucifixion study also collected hundreds of images of the resurrected Jesus from galleries around the world and asked, "Which cheek did the resurrected Jesus turn?"[33]

Unlike the 90 percent left-cheek bias found in images of the Crucifixion, pictures of the resurrected Jesus revealed a much weaker but still prominent leftward bias, with 49 percent of the representations featuring the left cheek, 21 percent portraying Jesus straight-on (see Fig. 30), and 30 percent featuring the right cheek.[34] Why is the left-cheek effect so much weaker in these scenes? Perhaps the "positive" emotion of the resurrection versus the Crucifixion is part of the answer. As we reviewed in the Introduction, the right hemisphere of the brain is dominant for emotional processing, especially for negative emotions. More positive emotions can be expressed by the left hemisphere of the brain, so the more "positive" scene of the resurrection could be reflected in the shift in the posing bias.

My own research group studied posing biases in religious art by comparing figures *across* religions.[35] Different religions have very different

Fig. 30: A centrally posed depiction of the resurrected Jesus.

approaches to emotional expression. Some contend that expressing strong emotions is important in everyday life and religious expression, a view associated with the Hebrew Bible and throughout the charismatic movement in Christianity.[36] The contemplative religious tradition takes a very different approach to emotion, in which the calming of emotions is a critical part of the religious experience. Given that Buddhism is an excellent example of the contemplative tradition, we chose to compare portrayals of Buddha to those of Jesus to see if the more emotional portrayals expected from Christianity would disappear in art depicting Buddha. As predicted, Buddha was much more likely to be portrayed centrally (no leftward bias; see Fig. 31) compared to representations of Jesus.

It might seem silly, even crazy to consider images of pets in this conversation of posing biases. After all, does a dog, cat, or frog actually pose for a

Fig. 31: A centrally posed image of Buddha.

picture when a camera is pointed at them? Despite a few amusing and prob-ably misleading exceptions on YouTube, pets probably don't know they are getting pictures taken, and it is even more unlikely that the right hemisphere of a dog is so emotionally dominant that the animal wants to present its left cheek during a particularly emotional portrait.

However, portraits of pets can be revealing of something else: the biases of the owner. Students in my own research group investigated this question by sampling pictures of dogs, cats, lizards, and fish, comparing all four non-human species to pictures of human babies.[37] Why babies? Because unlike the adults posing for a painting to be hung in the hall of the Royal Society, infants were unlikely to know the purpose of a photograph even if they understood what a camera and photograph were. Just like the dusty old portraits of dusty old men hanging in galleries around the world, the modern photographs of infants exhibited the same left-cheek bias. What about cats and dogs? Images of dogs tended to feature the left side of the face, but no such bias was observed for cats! What a surprise … cats do whatever they want. And fish and lizards? No bias.

TAKEAWAYS

So how will we pose for our next photographs, or which selfie will we choose for our next posts? If we want to appear emotional, approachable, and friendly, we should turn the left cheek. If we want to appear impassive, objective, and even aloof, we should pick a more central or even rightward pose. Sometimes posing "right" is posing left.

7

Lighting Biases: Do We Have the Right Lighting?

The substance of painting is light.

— ANDRÉ DERAIN

The famous cliché tells us that a picture is worth a thousand words. That's all well and good if we have a picture on hand. Putting pictures into words can be difficult. Using words to describe how our brains turn two mostly blurry and incomplete two-dimensional retinal images into seamless, clear, and cohesive three-dimensional representations is even harder. Our brains are great at constructing images, and a huge amount of neural real estate in our heads is dedicated to the job. However, the images that actually reach our retinas can often be interpreted in multiple ways. We don't tend to suffer through any of the alternatives, though. We perceive a visual scene in a single stable, cohesive way, instead of flipping between the plausible interpretations of ambiguous data. But that doesn't mean that what we "see" is actually there.

When describing his memory, Mark Twain famously said, "It isn't so astonishing, the number of things that I can remember, as the number of things I can remember that aren't so."[1] His musings about the fallibility of memory were ahead of his time, but Twain could have just as easily been talking about vision instead of memory. We "see" a lot of things that aren't actually there, and conversely, we fail to discern many things that actually *are* there. But we already know that. Anyone who has browsed through examples of optical illusions or watched a magic show quickly learns just how fallible our visual system is. It is a system that is easy to hack, and many of the calculations it makes when creating a visually cohesive world are based on some pretty simple assumptions.

Without these simple assumptions, it is very difficult to disambiguate the information coming in through our eyes. Consider the pair of spheres in Fig. 32. Which one is convex (sticking out) and which one is concave (pushing inward)?

Fig. 32: Two identical but rotated spheres. One appears convex, the other concave.

That is a trick question because both spheres are actually the exact same image, just rotated 180 degrees from each other. Neither one is concave or convex, but almost everyone perceives the leftward image as convex and the rightward one as concave. Why? Because of our assumptions. All things being equal, we assume the light source is normally from above.[2] This assumption was originally noted by Philip Friedrich Gmelin, a professor of botany and chemistry in the Royal Society of London, in 1744, and under normal circumstances, it holds up pretty well, with exceptions.[3] Unless we are on top of a mountain or looking at light reflected off water or snow, natural light usually comes from overhead. Even artificial light sources also tend to come from above. Most of us install the lights in our homes in the ceilings, not the floors. Outside our own homes (or planet), these assumptions don't always hold up.

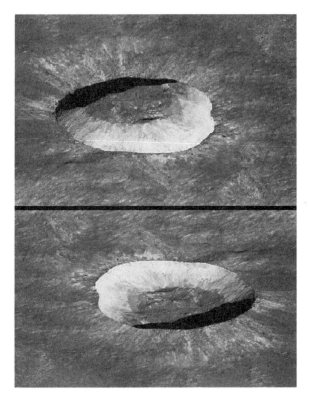

Fig. 33: A picture of a moon crater rotated 180 degrees. Assuming that the light source is from above makes the top image appear like a hill and the bottom one resemble a crater on the surface of the moon.

Consider the two copies of the same moon-crater image in Fig. 33. Assuming the light is coming from above makes the top image look like a crater, but the bottom one appears to depict a hill on the surface of the moon.

Shading is only one of the ways our brains use two-dimensional visual information to recover the third dimension. Other cues, such as interposition (one object occluding another), binocular disparity (the small differences in the images reaching the right versus the left eye), and linear perspective (the way linear perspectives converge in the distance) all give depth cues, too. Even the relative motion of two objects (called motion parallax) offers the viewer hints about the relative depth of objects. However, instead of providing a comprehensive survey on depth perception, most of this chapter will focus on light and shading, and specifically, on our biases and assumptions.

Learning that our visual system assumes that light comes from above probably shouldn't be a huge shock. After all, all species on Earth, even nocturnal animals or those living in the very dark depths of the ocean, evolved with an overhead light source. However, this book surveys leftward and rightward biases, so we should expect a side bias in lighting, as well, and indeed there is one. We tend to assume that the source of light isn't just from above but also from the left, and know this from field studies of famous artworks and lab studies of shaded bubbles such as the ones in Fig. 37. The leftward lighting bias can even be seen on old maps. Trying to create a sense of depth from a two-dimensional image is an old problem, and occasionally a really important one. For example, early cartographers had the challenge of depicting uneven terrain in a two-dimensional drawing, so they developed the custom of placing the shadow of a hill on the right side (inferring a leftward light source) as early as the 15th century (see Fig. 34).

When we go about our everyday lives, the light sources that illuminate our way are probably balanced between upper right and upper left. But if we survey famous paintings in Paris's Louvre, Madrid's Prado, and Pasadena, California's Norton Simon Museum, we find they tend to depict scenes or people lit from the upper left.[4] For example, consider *Allegory of Fortune* by Frans Francken (circa 1615) from the Louvre in Fig. 35. The source of the light is explicit and leftward (the sun in the top-left corner can be clearly seen). The light source is less obvious in other works and must be inferred by the shading.

Fig. 34: Cartographers developed a convention for portraying depth in two-dimensional images by shading changes in elevation with a presumed light source from the upper left. This is a modern digital example, but the convention emerged as early as the 15th century.

Fig. 35: *Allegory of Fortune* by Frans Francken (circa 1615) from the Louvre in Paris.

Jennifer Sun and Pietro Perona[5] studied 225 master paintings by giving two naive, independent raters (one left-handed, one right-handed) a protractor and asking them to determine the predominant lighting angle for each artwork. The bias for lighting from the top left was strong and consistent across time periods and artistic schools. It was present for Roman mosaics all the way through art from the Renaissance, Baroque, and Impressionist periods.

The same leftward bias in lighting direction is evident in some religious art.[6] In a study of Byzantine and Renaissance paintings of the Crucifixion and the Madonna and Child, the leftward bias was even stronger. In my own lab, we studied leftward lighting biases in art using a couple of different methods. First, we tried a really simple method for studying relatively simple art. We collected as many children's drawings as we could find online with an explicit light source. The sun had to be part of the picture. Once we had more than 500 distinct images, we coded them for lighting position. What did we find? Over two-thirds of the time (68 percent), children tended to draw the sun in the top-left corner of the page.

One of our subsequent studies was a little more sophisticated.[7] We wondered if the leftward lighting effect was present in *abstract* images created by grown-ups, paintings devoid of the typical elements found in portraits or landscapes. The light source in an abstract image can be difficult or even impossible to identify, so we had to come up with a clever method of looking for lighting biases. We created a "virtual flashlight" computer program controlled by a mouse and had people explore abstract paintings on a computer screen by illuminating the work, placing the spotlight wherever they wished. We instructed our participants to "position the virtual flashlight in a way that makes the painting most aesthetically pleasing to you." Just in case the abstract paintings we selected had lateral biases inherent in them already, such as putting the more interesting or attention-drawing elements in the upper-left quadrant, we presented each image normally and in a mirror-reversed orientation. Our participants viewed a randomized set of original and mirror-reversed images, and each time we displayed an abstract painting, they "lit" the painting in a way that looked best. What happened?

People chose to light to the top left even in these abstract images. On average, participants selected the upper-left quadrant for all but 6 of the 40 images. Because the leftward bias persisted for abstract art, we knew that the leftward biases in paintings of landscapes or portraits were not dependent on the discrete and concrete images in the work. Instead, a more fundamental bias had to be at work.

When looking at a painting, identifying the intended light source can be difficult. In some images (like *Allegory of Fortune*), the light source is explicit. We can see the sun in the upper-left corner. In other paintings, we need to infer the source given the patterns of shadows. For the most part, the studies we've considered have required people to judge where they think the light is coming from, but there's a problem with that methodology. As we have learned thus far in this book, humans have lopsided perceptions and behaviours, so it's entirely possible that these biases influence the results of these studies. This same flaw applies to the studies I have conducted myself. So how can we remove the potential confounding factors of a human rater?

We can do that by using a computer and some notoriously difficult-to-use image processing software called MATLAB (with apologies to MathWorks, Natick, Massachusetts, producers of MATLAB).

A research group with scientists from Japan and California devised a really clever study of frame composition in photography.[8] They had people

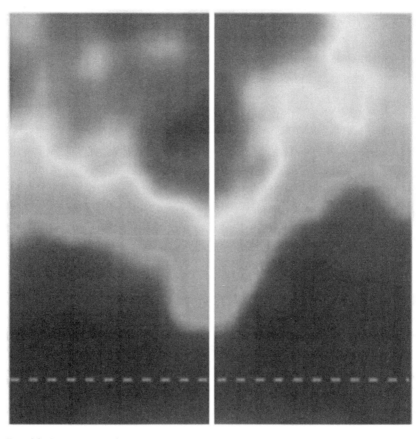

Fig. 36: An average of the lighting distribution biases across thousands of "composed" indoor photographs, demonstrating the bias toward lighting from the top left. The dotted line represents the horizon, while the vertical line represents the centre of the image. Note that the lighting source tends to be from above and left of true centre.

take more than 12,000 photographs across three conditions: (a) during the daytime outside with the instruction to *not* compose the frame of the photograph intentionally (the control condition in which participants simply rotated 45 degrees before taking the next photograph); (b) during the daytime outside with the instruction to frame the photograph intentionally; and (c) inside with artificial light sources with the instruction to frame the photograph. If people have a natural preference for images with the light source coming from the top left, that bias should emerge for conditions (b) and (c) but not for (a) where the framing of the photograph was essentially "random."

Instead of having humans judge the light source for the images, the researchers performed spectral analyses of the thousands of photographs, producing "average" images of the three conditions across all the photographs. Just as predicted, there was a consistent leftward tilt of the lighting gradients for the photographs for both of the "frame-composition" conditions (b and c), reaching 9 degrees leftward for the photographs taken inside (see Fig. 36).

Thus far we have considered images such as portraits or landscapes, even abstract art. Most of these pictures are pretty complicated, containing many different visual elements and colours, especially in the case of abstract works, and the elements are open to interpretation. However, our leftward lighting biases are obvious even with very simple images, ones that can only be interpreted in one or two ways. Look back at Fig. 32 with the two spheres, one that appears concave, the other convex. Those two images are 180 degrees apart, but what if we didn't rotate them quite that much? What if the light didn't appear to be coming directly from the top or bottom, but from the side instead?

Jennifer Sun and Pietro Perona[9] tried just that using clusters of shaded spheres (see Fig. 37) and asked people to detect a single target bubble that didn't match the rest of the cluster. They expected people would be fastest at detecting the "oddball" bubble when the overall light source was depicted from directly above, but that was not what happened. Instead, people were best able to detect the oddball when the light was coming from the top left by about 30 degrees from centre. Other researchers employing a slightly different method (images of flat surfaces with parallel protruding strips)

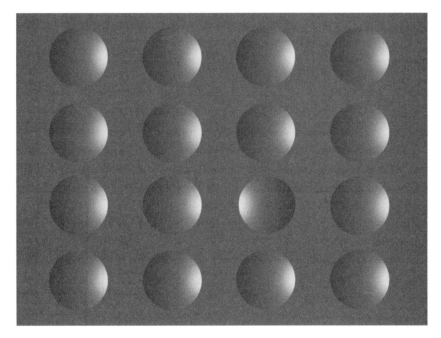

Fig. 37: A cluster of bubbles mostly lit from one direction (from the right in this case). People had to detect the "oddball" bubble that doesn't match the rest of the cluster.

found almost the same thing.[10] In that study, people preferred lighting that was 26 degrees to the left.

In my own lab, we tried something very simple, inspired by Fig. 32. If the sphere appeared to be convex when lit from the top and concave when illuminated from the bottom, what would happen if the light came from the side? We took pairs of spheres and varied the lighting angle by 22.5 degrees, then asked people to judge which sphere seemed concave. Just as one would expect from the surveys of famous paintings, people showed a bias toward images lit from the left. Leftward-lit spheres were convex (as if they were illuminated from above), and rightward-lit images were concave.

There are many other studies of clusters of bubbles I could mention, but the vast majority, though not all,[11] find the same lighting bias I've already described.[12] Some observe that factors such as handedness and head tilt modulate but don't reverse or eliminate the bias. Objects lit from the left even appear "brighter" than the same but mirror-reversed image lit from the right.[13] Studies of "real" three-dimensional art, such as sculpture, are few and far between and so far have yielded mixed results.[14,15]

Almost all the studies we have considered up to now have come from the Western world. Western portraits, art, and advertisements perceived by Western people clearly have leftward lighting biases. However, as we learned in Chapter 3, an individual's native reading direction (NRD) can have a big influence on lopsided everyday behaviours. The text in this book is written in English and is scanned from left to right as it is read. Most modern languages of Europe, North and South America, India, and Southeast Asia are written from left to right. However, there are some popular languages that read from right to left, including Arabic, Aramaic, Hebrew, Persian/Farsi, and Urdu. For readers who have a native language that reads right to left (RTL), might lighting biases look different than those from the West who read left to right (LTR)?

We can also answer this question by using the convex/concave bubbles as before. If we make an array of the bubbles (devising an image that looks kind of like an egg carton) but "light" one of the bubbles from the "wrong" (i.e., different) angle, that "oddball" stands out (or in). Consider the array in Fig. 37 again. The bubble in the third row and third column is the oddball. Based on most of the studies to date, we are considerably faster at detecting oddballs lit from the top left than those from the top right. However, there are exceptions. Hebrew is read from right to left, and Hebrew readers typically show reduced leftward lighting biases or even rightward biases.[16,17]

Given this reversal in the bias, we wondered if a person's *preference* for objects lit from the left or right might also change with NRD. In my own lab, we presented images lit from one side or the other along with a mirror-image copy of the same scene (see Fig. 38) to LTR and RTL readers. In addition to asking people to tell us which image they preferred, we also used an infrared eye monitoring device to measure where people were looking on the screen while they compared the images.

People accustomed to reading from left to right spent much more time studying the left part of the screen than the right, and they also preferred the images that were lit from the left. Right-to-left readers didn't quite show the opposite pattern but came close. They spent a great deal more time scanning the right side of the images instead of the left, and the preference bias toward the left-lit images went away but didn't reverse direction.

So far we have learned that most people (1) tend to assume light comes from the top left; (2) are inclined to produce art with a leftward lighting bias; (3) respond faster to objects lit from the left; (4) think that leftward-lit objects seem brighter; and (5) prefer objects lit from the left. But how do these lopsided perceptions influence us in the real world?

The leftward lighting bias we see in old paintings is also evident in modern-day print advertisements. In a survey of 2,801 full-page ads, my own research group found that 47 percent of the ads were lit from the left, 33 percent from the right, and 21 percent exhibited central lighting.[18] In the previous chapter, we discussed the posing biases commonly found in portraits and ads, with most people showing the left cheek in both. It should come as no surprise that posing and lighting biases influence one another. Ads featuring a rightward pose also tend to be lit from the right, and vice versa.

Try as I might, I have not been able to find any proof that artists or advertisers are actually doing this on purpose. The evidence of their lopsided creations is very clear, but why are they that way, and in the case of ads, does lopsidedness actually help? Are ads lit from the left any better?

Based on some preliminary work out of my own lab, it looks as if they are better. Knowing that most ads are lit from the left, and that there might be some other underlying differences in layout or content between left- and right-lit ads, we sought to fashion our own fake ads for fictitious products, making equivalent left-lit and right-lit versions of the same ad (see Fig. 38). We created fictitious product names to avoid influence by attitudes toward real brands, presented both versions of the ads to 45 students, and asked them to rate things such as their attitudes toward the advertisements, attitudes toward the products, attitudes toward the brands, and their willingness to purchase the items in the future. Our phony consumers gave higher ratings to bogus products and brands when they were lit from the left. It appears as though the leftward bias we see in real ads pays off, after all.

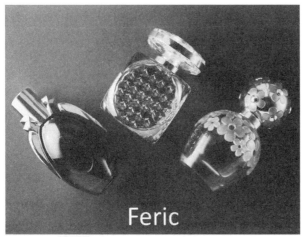

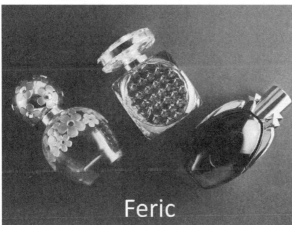

Fig. 38: Fake advertisements with both left-lit and right-lit versions. We asked people which product they'd prefer to buy; they chose ones lit from the left.

TAKEAWAYS

In this chapter, we learned that our visual perception is guided by our assumptions, and that we tend to assume that light sources come from above and to the left. This bias is reflected in our portraits, art, and even advertisements. We think that objects lit from the left look better, and we're even more likely to buy them. The next time when posing for a profile picture for an online dating service, or perhaps taking a snapshot of that old couch to sell online, it might help to remember to keep the bright side on the left.

8

Side Effects in Art, Aesthetics, and Architecture

The right eye is the counsellor which shows the
way in divine things; the left eye shows the way in
worldly things.
— ST. AUGUSTINE OF HIPPO (354–430 CE), *On the
Lord's Sermon on the Mount*

De gustibus non est disputandum is a Latin maxim that technically translates to "In matters of taste, it should not be disputed/discussed." The
more common English interpretation of the phrase is "There is no accounting
for taste," which reflects the widely held belief that taste in art and aesthetics
is highly individualized and often unpredictable. A more extreme interpretation of this maxim is one in which science in general (and experimental
psychology in particular) is of no relevance to aesthetics.[1-4] It's true that the

science of art is still in its childhood,[5] but we are making progress. In this chapter, the science of laterality barges its way into the art world.

As we will learn, there are some clear and consistent lateral biases in the production and appreciation of art and architecture. We will also discover that an individual's aesthetic preferences can be quite complicated and influenced by a multitude of factors. Laterality is one of those factors, but others can certainly be more important and even override any influence lateral biases might have in a given situation. Simply mirror-reversing a genuinely terrible painting will not transform it into a beautiful or even acceptable piece. However, under some circumstances, changing the orientation or even the lighting in a work can make it better or worse from the observer's point of view.

Much of art is lopsided, and we have already examined several important examples of these biases. Painted portraits tend to portray the left cheek of the poser. Pictures of a mother holding a child, such as Mary clasping Jesus, tend to depict the child cradled to the left. The light source in paintings or photographs usually emanates from the top left. Of course, not all art is lopsided. Humans also find symmetry to be aesthetically pleasing. It's one of the things we often look for in an attractive face or even a beautiful building. Consider the bilateral symmetry in the Cathedral and Metropolitical Church of St. Peter (commonly known as York Minster) in York, England, in Fig. 39, or the radial symmetry in the Bahá'í House of Worship (Lotus Temple) in Delhi, India, in Fig. 40.

In his book *Symmetry*,[6] noted German physicist, mathematician, and philosopher Hermann Weyl explicitly links beauty with symmetry: "[In the one sense] symmetric means something like well-proportioned, well-balanced, and symmetry denotes that sort of concordance of several parts by which they integrate into a whole. *Beauty* is bound up in symmetry [emphasis in original, as quoted in McManus]."[7] [...] "Some early thinkers such as Aristotle thought beauty was a property that arose from symmetry, but other scholars like Plotinus thought symmetry and beauty were independent of one another."[8]

Symmetry in art isn't limited to a single modality. One can appreciate the symmetry in a painting, a cathedral, or even a sonata (with an A-B-A

Fig. 39: An example of bilateral symmetry (lack of lateral bias) in the Cathedral and Metropolitical Church of St. Peter, commonly known as York Minster, in York, England.

Fig. 40: An example of radial symmetry in the Bahá'í House of Worship (Lotus Temple) in Delhi, India.

Fig. 41: A self-portrait etching by Rembrandt.

musical form).[9] Considerable value is placed in symmetry in art, and despite my focus on lopsided art for the remainder of this chapter, I don't want to leave the reader with the impression that all great art is asymmetrical, or that making art more asymmetrical will cause it to be more appealing. However, there are many examples of systematic, predictable asymmetry in the art world, such as the examples of the posing, lighting, and cradling biases we have already explored. These biases are clearly related to the underlying asymmetries in our brains.

Discerning the left from the right half of a piece of art might seem like a simple task, but in many cases it is anything but. Consider one of Rembrandt's etchings, the self-portrait seen in Fig. 41. Which is the right side? An etching serves as a method for making prints, pressing the etched plate to a sheet of paper with a pigment in the middle. In fact, some art historians claim that Rembrandt's etchings can only be studied properly by examining their mirror-reversed image, because the printed product was the visual element Rembrandt was intending to share. However, the image in Fig. 41 is even more confusing than that. It's a self-portrait, after all. Was Rembrandt making this work based on a reflection? Is the etching a mirror reversal of an already mirror-reversed image? This challenge of discerning right from left isn't unique to older artworks. The same problem applies to modern selfies taken with a smartphone. In many cases, the camera/phone is clearly visible in the image, and the observer can be sure that the selfie was taken in a mirror. In other cases, it can be much harder to discern, and whether it is a "mirror" selfie or not needs to be inferred by looking at text in the background of the image, or unique, perhaps asymmetrical facial features.

Aestheticians such as Mercedes Gaffron[10] and Heinrich Wölfflin[11] have argued that the left and right halves of a piece of art have distinct meanings and that mirror-reversing a work alters its meaning. Geometrically, two

Fig. 42: *Reading Woman* by Pieter Janssens Elinga.

mirror-reversed pictures contain all the same elements. However, the perceptual experience of visually exploring two "equivalent" but mirror-reversed images can be quite different. Consider the example of *Reading Woman* by Pieter Janssens Elinga in Fig. 42. On the left is the original image, with the mirror-reversed reproduction on the right. In the original painting, the woman appears to be more salient, perhaps because she's viewed "earlier" in the scan of the painting. The woman's slippers on the floor might not stand out when viewing the original, but in the mirror-reversed picture they probably appear out of proportion and rather "in the way." Even the angles of the floorboards seem quite different in the two versions.[12] Some artists, such as Vincent van Gogh and Albrecht Dürer, took great care to ensure viewers were exposed to prints in the correct orientation, whereas others, like Raphael and Edvard Munch, did not.[13]

In addition to the aforementioned rather subjective accounts of the different interpretations of mirror-reversed versions of the same image, there are more objective ways to demonstrate a similar effect. For example, we can use optical illusions. Look at Fig. 43 to see the "corridor illusion" first reported in 1950 by American psychologist James J. Gibson in his article "The Perception of Visual Surfaces."

The two bars are actually the exact same size, but the one that appears farther down the corridor seems larger. Now, instead of placing the bars in the middle of the image, position them on one side or the other of the corridor and measure the strength of the illusion. A clever study by Samy Rima and colleagues[14] employed mirror-reversed corridors and tested left-to-right reading viewers from France and right-to-left ones from Syria. The illusion was strongest for left-right readers when the corridor was along the right side of the image, but stronger for right-left readers when the corridor was along the left side. Not only does this experiment empirically demonstrate the qualitatively different experience between the two mirror images, but it shows how it reverses with native reading direction.

Native reading direction also appears to influence anisotropy, a property in art of being directionally dependent. Different directions have different properties, as compared with isotropy, in which the different directions are functionally equivalent. One common element of visual anisotropy in painting is the "glance curve" described by Heinrich Wölfflin[15] in which

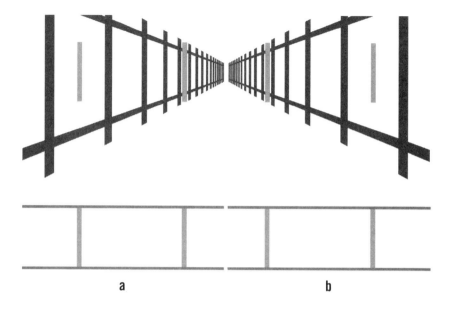

a   b

Fig. 43: The "corridor illusion" first reported by James J. Gibson in 1950. In both (a) and (b), the bar depicted farther down the "corridor" appears to be larger/longer than the one closer to the viewer, but the depiction below proves they're indeed the same length. When presented to groups that differed in natural reading direction, left-to-right readers (from France) experienced a stronger illusion when the corridor was along the right side of the image (b), whereas right-to-left readers (from Syria) experienced a stronger illusion for corridors along the left side (a).

individuals begin visual exploration in the lower left portion of a work, progressing upward and to the right side. For Western artists and observers, movement from left to right is "easier and faster," whereas movement from right to left is "slower and perceived as having to overcome resistance."[16] These lower-left to upper-right vectors are readily observable in many Western art pieces (see Fig. 44). We tend to perceive the diagonal from bottom left to top right as ascending, whereas we perceive the anisotropic diagonal (from the top left to the bottom right) as descending.[17,18]

Fig. 44: An example of perceiving the anisotropic diagonal as rising from the bottom left to the upper right.

Although some researchers[19,20] have argued that the lower-left to upper-right visual exploration vector is a fundamental bias exhibited by everyone, we can easily find exceptions from cultures with languages that read from right to left. For example, ancient Chinese handscrolls open and read from right to left. Images in these scrolls typically include the main visual element biased to the right side, and the direction of the implied movement from the figure is also right to left.

We can also observe these rules of directional motion in live theatre. In the West, stage right (the audience's left) is more likely to draw the audience's attention, thus when the curtain rises for the beginning of an act, people tend to look to their left for the beginning of the action.[21] This convention is reversed in Chinese theatre. The most important positions are stage left (the audience's right), congruent with the reversal in reading direction between Western and Chinese audiences. German psychologist Mercedes Gaffron has claimed that we "read" visual scenes in much the same manner as we read books, and there appears to be mounting evidence to support that claim.

Fig. 45: Sample linear orderings of family photographs. In countries with left-to-right reading direction such as Spain, family portraits tend to ascend from left to right (leftward cluster). Conversely, family portraits from Iran (with right-to-left reading direction) usually ascend from right to left (rightward cluster).

For instance, photography historian and academic Carmen Pérez González assembled and analyzed two large samples of 19th-century photographic portraits from two different countries,[22] Eight hundred and ninety-eight of the photos were from Spain (Spanish is written from left to right), whereas the other 735 were taken in Iran (Persian is written from right to left). She sampled five different types of images from these two regions and analyzed their directional biases. There were linear orderings of a group of people (often families) arranged by height; couples, in which one person stood and the other sat; a single person resting an arm on a chair; a single person seated and resting an arm on a table; and portraits with a single person posing without chairs, tables, or other props. Comparing across these categories, we can see that reading direction had a profound effect on the composition of the photographs. For the linear orderings and couples images, the images ascended in height in the same direction as the written word (see Fig. 45).

The images with chairs, tables, and even solo portraits also demonstrated clear influences of reading direction. In a follow-up study,[23] the same images were presented to Spanish (left-to-right reading) and Moroccan (right-to-left reading) observers in either the original or mirror-imaged orientation. Spaniards preferred the rightward versions of the photos, whereas the Moroccans favoured the leftward ones. Therefore, both the original image composition and subsequent choices for the same photo sets were clearly modulated by native reading direction.

The directionality in González's images was created by "ordering of mass." There was no actual movement in the images themselves, nor was any movement implied. Consider an "action shot" of someone walking from left to right. Even with a static image, the direction of implied movement is obvious, provided we do not imagine Michael Jackson performing the moonwalk in which the direction of implied motion and the actual direction of motion are opposites. Images of cars, trains, planes, cats, dogs, and pretty much anything mobile with a clearly discernible front and back can imply movement in one direction or another. Left-to-right readers appear to select images with implied left-to-right directionality, but the results with right-to-left readers have been less clear. Some studies find a reversal in the directional effect, whereas others discern none at all.[24,25]

This effect isn't limited to pictures of people. In my own lab, we demonstrated some of the same biases using images of mobile objects and landscapes.[26] We presented mirror-imaged pairs depicting motion from left to right or right to left to two groups of people: left-right readers and right-left readers. Although our right-left readers didn't show any clear preference for the directionality in either image type, the left-right readers had a strong bias toward left-right directionality in the images we presented to them. In addition to showing still images with implied movement in them, we also exhibited short video clips of directional movement to the same groups. In the video condition, the directional biases became even stronger. Left-right readers had even more pronounced preferences for videos depicting left-right movement, whereas right-left readers, who didn't display a clear directional preference for still images, now demonstrated a stronger bias for right-left movement.

Employing the same sets of images and video clips of movement in objects or landscapes, we also looked at responses by non-Western viewers.[27] Comparing Hindi (left-to-right) readers with Urdu (right-to-left) readers, we found that the left-right ones had the same strong left-right bias as the Western sample did in the previous study, but the right-left readers didn't show much directional bias at all. This study taught us that the directional aesthetic biases exhibited by Western samples are not unique to the West but are also evident in other left-to-right reading cultures. As we will see in Chapter 12, these directional biases also have implications for aesthetic judgments in sports such as gymnastics.

Images with sequences of figures also reveal the Western observer's inclination for left-to-right directionality. Marilyn Freimuth and Seymour Wapner's 1979 study on directionality in paintings[28] presented works (in both original and mirror-imaged orientations) with sequences of figures, varied the location of the principal figure (left or right side of the image), and measured the aesthetic preferences of the observer. The location of the principal figure was not the primary determinant of preference. Instead, it was the sequence of figures that mattered most. People chose paintings that depicted left-right directionality, regardless of whether the image was presented in its original orientation or mirror-reversed. The Western favouring of left-right ordering even extended to the titles of artistic works. If the first word referred to content on the left side of the image, that title was preferred.[29]

In addition to reading direction, other factors can also determine lateral biases in art. Consider the example of drawing a facial profile. Barry Jensen[30,31] at the University of Kentucky asked right-handed and left-handed people from the United States (left-to-right language), Norway (left-to-right language), Egypt (right-to-left language), and Japan (normally a right-to-left language) to draw facial profiles. Regardless of the native reading direction, the right-handed artists drew facial profiles that faced leftward, whereas left-handed artists didn't display consistent lateral preferences at all. More recent studies[32,33] have replicated and extended these findings, adding conditions in which people had to draw trees, hands, even fish. In these additional conditions, both the handedness and the native reading direction influenced the orientation of the drawings.[34]

Of course, the way artists organize visual scenes is not strictly driven by aesthetics. Not all artists are simply trying to make a picture as "pretty" as possible. Instead, artists typically try to communicate something and evoke an emotion with an image. Subtleties in colour of tone, relative position, and even texture can all shape the intended message in a painting. As it turns out, the lateral biases we see in images also depend on the intended message, as well as implied relationships between the objects and actors in the picture.

Consider a very simple and elegant experiment by Anjan Chatterjee and colleagues at the University of Pennsylvania.[35] They asked people to draw an agent/receiver relationship such as "the circle pushes the square." In that example, the "agent of action" was the circle, whereas the "receiver of action" was the square. Chatterjee and his team found that people tended to draw that picture with the circle (agent of action) on the left. The researchers suggested that the left-cheek bias in portraiture could be explained by a leftward agency bias. If painters wanted to portray the agent of action to the left, the agent exposed more of the right cheek. Because females were typically depicted as more passive, they should be less likely to be depicted as the agent of action, resulting in a stronger leftward bias.[36]

In Western cultures, images of a man and woman typically position the woman to the right of the man.[37,38] In a study of online images of Adam and Eve (acquired by searching for "Adam and Eve" or "Eve and Adam"), Eve was portrayed to Adam's right 62 percent of the time.[39] This gender bias has been termed the spatial agency bias by psychology researchers Caterina Suitner and Anne Maass,[40,41] and it is clearly visible in cultures that read from left to right, or with languages in which the subject typically precedes the object in a sentence.[42]

In a clever study of the spatial agency bias and how it relates to posing biases, Mara Mazzurega and colleagues[43] presented male or female faces in leftward or rightward orientations and had participants choose whether the depicted individual had a high-agency job (stockbroker, architect, lawyer, chef, engineer, film director) or low-agency employment (flight attendant, secretary, mail carrier, call centre worker). Right profiles were perceived as gender atypical for females and less associated with the high-agency jobs (see Fig. 46). In other words, leftward profiles were seen as gender typical for males and more associated with agentic occupations.

THE FOUR COMBINATIONS OF GENDER AND FACE ORIENTATION

Female    Male

AMBIVALENT SEXISM SCALE

Gender stereotype-congruent spatial association task:
Is the face oriented to the left or to the right?

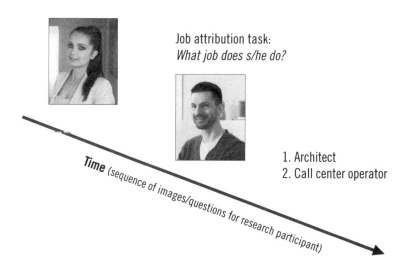

Job attribution task:
What job does s/he do?

1. Architect
2. Call center operator

Time (sequence of images/questions for research participant)

Fig. 46: Schematic diagram of the method used by Mara Mazzurega to detect how the spatial agency bias relates to posing biases. After presenting male or female faces facing rightward or leftward, participants had to indicate whether the presented face fit with a high-agency job (stockbroker, lawyer, et cetera) or low-agency employment (flight attendant, mail carrier, et cetera). Rightward profiles were associated with female faces and low-agency jobs, whereas leftward profiles were associated with male faces and high-agency employment.

One of our grandest art forms is architecture, which can be lopsided in a couple of interesting ways. Earlier in this chapter, I discussed symmetry (both bilateral and radial symmetry) and mentioned the symmetrical Bahá'í House of Worship (Lotus Temple). However, architecture can also exhibit chirality, the object property of having distinct right and left forms. Chiral objects such as wood screws are common in everyday life, while particles as small as atoms can exhibit chirality, as can very large things such as skyscrapers or even galaxies.[44]

Many famous buildings are achiral, meaning they exhibit mirror symmetry and don't have distinct left and right forms (Egyptian Pyramids, Taj

Fig. 47: Examples of chirality in architecture. The Mode Gakuen tower in Nagoya, Japan, on the left, turns counter-clockwise, whereas the Turning Torso skyscraper in Malmö, Sweden, on the right, twists clockwise.

Fig. 48: Mirror-imaged historiated columns at the bilaterally symmetrical Church of St. Charles Borromeo in Vienna.

Mahal, Empire State Building). Numerous examples of great architecture, however, aren't symmetrical at all, such as the Guggenheim Museum in New York City, the Harpa Concert Hall in Reykjavik, Iceland, or the Perot Museum of Nature and Science in Dallas, Texas.

To be sure, those asymmetrical examples of achiral buildings are pretty extreme. Another common source of achirality in architecture comes from spiral elements such as those found in the leftward spiral of the Turning Torso skyscraper in Malmö, Sweden, or the rightward spiral of the Mode Gakuen tower in Nagoya, Japan.

The directionality in spirals in architecture has been studied extensively. Spiral columns — often with congruent spiral staircases inside — are common throughout the world. Trajan's Column in Rome (circa 113 CE) was an unprecedented monument in its time and inspired a number of other similarly designed, rightward-turning, columns.[45] The rightward turn of the spirals within the column was congruent with the screw designs used by drafters back then and was also consistent with the left-right directionality of written Latin.

However, most Greek and Roman art depicts the motion of figures from left to right.[46] When German art historian and archaeologist Heinz Luschey first made this observation, he rejected the possibility that the lateral bias was related to the directionality of the written word, because Egyptian art from the same period demonstrated the same left-right directionality, and many of the Greek images showing the bias were created before the directionality in Greek script had been firmly established.[47] The vast majority of historiated columns wind upward and to the right. There are some exceptions, such as the leftward spiral on the Bernward Column in Hildesheim Cathedral in Germany, and in some cases, leftward and rightward spiral columns are strategically created to maintain bilateral symmetry, such as those in the Church of St. Charles Borromeo in Vienna (see Fig. 48).

TAKEAWAYS

There is no accounting for taste — *De gustibus non est disputandum.* Individuals have widely varying aesthetic preferences, and we struggle even to define concepts like aesthetics or beauty in a meaningful and cross-cultural way. It is very hard to come up with an objective definition and method of measuring aesthetic experiences when they are subjective and multifactorial. However, there are clear lateral biases in the art we produce, as well as reliable lateral biases in how we perceive and react to it. Considerations such as posing direction, lighting direction, centre of mass, direction of movement, and native reading direction of the primary audience should all be taken into account when crafting a painting, plating a meal, or designing a skyscraper. Left-to-right readers prefer left-to-right motion (or even *implied* motion). This effect is reflected in how visual elements are arranged in Western art, how family portraits are typically arranged, and even how optical illusions are perceived. We can take advantage of these biases when creating images. If we live in a country that primarily reads from left to right and take pictures of sports cars with the intention of listing them for sale, consider positioning the vehicles as if they were moving from the left to right, positioning the light source to the left. Similarly, when arranging multiple elements in an image, arranging the objects with a left-to-right directionality will be more aesthetically pleasing and familiar for left-to-right readers.

9

Gestures: Leftover Behavioural Fossils

Why do people always gesture with their hands
while they talk on the phone?

JONATHAN CARROLL

Try to keep your hands perfectly still while you talk. It takes effort. Every time I teach a class or give a presentation about gestures during speech, I get quite self-conscious about my own hand gestures and even try to control them. It doesn't last long, though. Kind of like thinking about your breathing; you can consciously take control of the behaviour for a while, but sooner or later you return to your normal state of automatic breath/gesture control. Waving our hands around during conversation is both natural and unconscious. We can kid ourselves about how we might do it for the benefit of the other person, but watch any example of "disembodied" communication (telephone, walkie-talkie, hosting a radio show) and the notion is immediately undermined. Even when standing alone in a sound booth while recording a speech, our hands and arms are just as animated as they are while we converse face to face.

Not only do gestures persist in the absence of vision; they also appear in people who have never seen a hand gesture in their lives. Blind toddlers spontaneously gesture as they learn to speak,[1] although their gestures are less frequent than those produced by sighted children. Furthermore, people of all linguistic and cultural backgrounds gesture, so clearly there's something fundamentally communicative about them.[2] Gestures don't require an observer or even a model to teach children how to gesture. Speech and gestures are clearly linked, but how and why? What can our gestures teach us about our brains?

In the 1860s, Paul Broca, a French physician and anthropologist, studied a patient with a large injury in the left frontal lobe.[3] The patient's name was Louis Victor Leborgne, but he is commonly referred to as "Tan" because that became the only word he could speak and he used it in different tones of voice to communicate different things. After Tan died, examination of his brain revealed an absolutely massive brain injury in the left frontal lobe, an area now referred to as Broca's area (see Fig. 49).

Based on his examination of Tan, Broca concluded that the left hemisphere of the brain contained the speech centre, thus starting the speculation that continues to this day whether there is a functional connection between the left frontal lobe's dominance for speech and its coordination of the usually dominant right hand.

It's entirely possible that our species communicated through gestures long before we mastered the spoken word. Several prominent theories of human language development focus on a potential transition between communicating through gestures to communication through sounds.[4-6] However, the relationship between gestural language and spoken language still isn't well understood.

Regardless of whether gesture evolved before the spoken word or whether one led to the other, it is clear that we gesture *as* we converse. Not only do we gesture while speaking, we also gesture while listening. It's also apparent that when and how we gesture reflects the left-right differences in our brains.

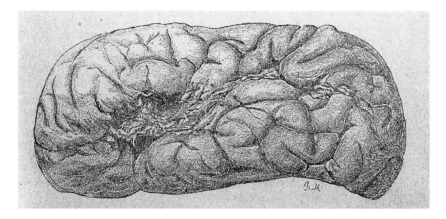

Fig. 49: Drawing of Louis Victor Leborgne's (Tan's) brain, showing a large lesion in the frontal lobe, a region now known as Broca's area.

In 1973, the Canadian psychologist Doreen Kimura had pairs of people who didn't know one another "fake" a conversation while being observed in the laboratory.[7,8] She coded the hand movements during speaking and listening, finding that (1) people moved their hands more while speaking than listening; (2) they gestured more with the right hand than the left while speaking; and (3) they gestured more with the left hand while listening. All three effects were evident in right- and left-handers, and Kimura concluded that the right-hand movements during speech were due to the left hemisphere's dominance for speech production, and that the same underlying brain circuitry that underpinned speech was responsible for the increase in right-hand movements.

This finding was replicated and extended by John Thomas Dalby and colleagues.[9] Instead of manufacturing conversation between strangers in the laboratory, Dalby's study examined real conversations, "in the wild," between pairs of people who knew one another. Just as Kimura discovered seven years earlier, people tended to perform free movements with the right hand during speaking but not while listening.

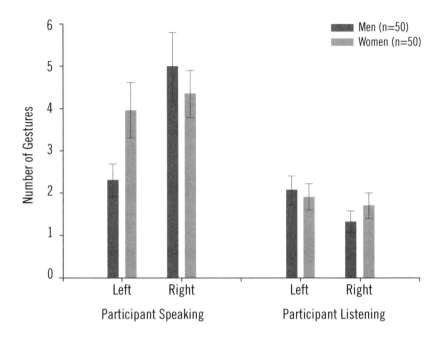

Fig. 50: People tend to gesture with their right hands while speaking, and this left-right difference is stronger in males than females.

My own research group extended these studies by searching for male-female differences during natural conversations between pairs of people.[10] We observed 100 conversations for three-minute periods for 50 males conversing with another male (25) or a female (25), and 50 females conversing with another female (25) or a male (25). We coded the gestures according to whether the person was speaking or listening, and whether the gesture was a "free movement" or a "self-touching" gesture (see Fig. 50).

The males we observed made more right-hand movements while talking, but while listening, males gestured with the left hand. The females we observed didn't quite demonstrate the same pattern. The left/right differences were not nearly as large for the females in either condition. We generally find this in other areas of laterality research, too. Males tend to show larger laterality effects than females.

Hand gestures can also be studied in people who use them as the primary means of communication — the deaf. For the "non-linguistic" (i.e., movements not corresponding to sign language) hand gestures, right-handed deaf signers gesture more with their right hands and vice versa for left-handers.[11]

Why does talking result in "overflow movement" of the right hand? It's certainly true that control of the right hand and dominance for language both tend to be concentrated in the left hemisphere of the brain, but the reason is probably even more specific than that. If we look at the image of the "motor homunculus" in Fig. 51 (homunculus translates to mean "little

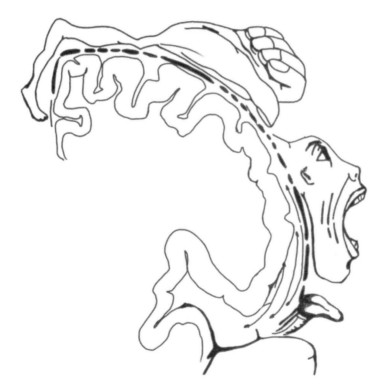

Fig. 51: The motor homunculus. In the part of the frontal lobe of the brain that controls bodily movement, note that control of the hand and mouth are right next to each other.

man"), the distorted proportions of the body parts and their representation in the cortex are probably what jumps out. After all, the hands and mouth and tongue are *huge*, whereas the leg and trunk are tiny. However, that isn't the only distortion that should stand out. The relative position of the body parts is wonky, too. The hand is next to the face, which is weird, but consider a scenario of "spreading activation" or "motor overflow" in which a lot of brain activity in one area tends to excite some adjacent areas. Lots of excitation of the brain tissue driving the mouth could also trigger some hand movement and vice versa. Perhaps speech and gesture go hand in hand because of the neighbouring brain tissue driving these two regions.

So far our discussion of gesturing during speech has ignored what people actually say. In many cases, as in the study I did on the topic, we didn't actually know what they were saying because we couldn't hear them! We observed gestures during natural conversations but were simply too far away to overhear what the pair had to say.

However, it looks as if the content of the conversation matters. In the course of most exchanges, it is the left hemisphere that dominates the content and subsequently waves the right hand about while someone is speaking. However, what happens if the topic shifts to something in which the right hemisphere is the expert? What if people are giving directions or conveying other spatial information?

This intriguing question was asked and answered by Sotaro Kita and Hedda Lausberg,[12] who studied a small group of patients that had undergone a very special brain surgery — the severing of the corpus callosum, the "bridge" between the two hemispheres of the brain. Except for some very rare cases of people born without a corpus callosum,[13] most people have a large band of white matter tracts (around 250 million of them) that connect the two hemispheres. Of course, the two halves of the brain normally collaborate to collectively form perceptions and actions.

In some rare and severe cases of epilepsy, physicians resort to the rather extreme measure of disconnecting the two halves of the brain in an effort to keep seizures from spreading from one spot on one side of the brain (usually in the left temporal lobe) to the whole brain. This procedure of cutting the corpus callosum is called a callosotomy.[14] Following the procedure, the side effects tend to be quite subtle. The two halves of the brain continue to perform their specialized tasks but with less collaboration with each other.

In the Kita and Lausberg study of hand gestures, three patients with complete callosotomies were compared to nine neurotypical control participants. Two of the three patients had left hemisphere dominance for language, but the third had some language processing in both hemispheres of the brain. All three patients with the callosotomy produced spatial imagery gestures with left and right hands, but the two with left hemispheric language had trouble gesturing with the left hand to give spatial information along with the spoken content. This suggests that the right hemisphere alone can generate speech gestures about spatial content, just as the left hemisphere can, which is in sharp contrast to the earlier claims by Doreen Kimura and others that the same brain areas responsible for speech also generate our hand gestures as we speak. That might be the case some of the time, but when describing content that taps into the expertise of the right hemisphere, it appears that the right hemisphere can also generate its own gestures.

Further evidence that "the message matters" emerged from a study of 10 right-handed neurotypical adults chosen from a cohort of 122 healthy adults.[15] These participants were presented with animations and had to give verbal or silent gestural demonstrations of the animations. People used different hands to gesture depending on the content of the animations. In the verbal condition, people initially preferred the right hand for making gestures. For describing a scene, the hand employed to gesture about the object tended to correspond to the location (left or right) of the object. This resulted in an "iconicity" of the gestures in which the gestures themselves were closely related to the meaning of the spoken content. When referring to objects in the left part of the animation scene, people resorted to their left hands. There was no difference between the speech and silent condition in the laterality of the gestures.

The left hemisphere is typically the "linguistic" expert, in that it knows more words, understands the consequences to meaning driven by word order (grammar), and can produce the spoken word. However, there are language-related tasks in which the right hemisphere dominates. It is clearly better at decoding tone of voice (everything from emotion in speech to sarcasm), extracting a theme from a narrative, and even understanding metaphors.

This superiority at metaphor reveals itself though hand gestures, too. When 32 English speakers were required to gesture with either the right or left hand to indicate a metaphor, such as "to spill the beans," the metaphor explanations were superior while gesturing with the left hand.[16] Furthermore, when people engaged in either left- or right-hand gestures, the amount of metaphoricity (frequency of using metaphors) increased during left-hand gesturing.

I want to end this discussion of lateral biases in gesture with an intriguing idea. What if our hand gestures during speech are evolutionary relics, a behavioural fossil left over after humans developed spoken language? We tend to think of fossils as petrified bones, of course, not behaviours. However, modern humans are rife with behavioural fossils, tendencies that were adaptive during the bulk of our time on Earth (especially the 100,000 years of human evolution on the African savannah), but these behaviours aren't necessarily adaptive or even useful today. At worst, some of them are actually maladaptive today. Consider the universal human behaviour of craving sweet foods. On the African savannah, our distant ancestors who preferred sweets would have consumed more fruits and enjoyed the vitamins and other nutritional benefits therein. There were no candy stores on the savannah back then. Choosing sweets was a healthy choice. Fast-forward thousands of years to today, and sweets are everywhere (especially in the West), and almost never nutritious. The sweet tooth that might have helped keep our ancestors alive in Africa could be actively harming our health today.

The idea that manual gestures are behavioural fossils was advanced by the French philosopher Étienne Bonnot de Condillac in 1746, but more recently furthered by American Gordon Hewes in 1973[17,18] and especially by New Zealander Michael Corballis later.[19] Remember the famous case of Tan described earlier in this chapter? Following damage to part of his frontal lobe (Broca's area), Tan lost his ability to articulate words other than *Tan*. Broca's area in the human brain corresponds to area F5 in the monkey, a region tasked with controlling manual gesture (not vocalizations).[20] Furthermore, if we record single brain cells in a monkey's F5, those cells appear to be part of the "mirror system" in the primate brain. These special cells are responsive to the animal making a reaching movement toward an object *or* even if the monkey views another member of the group reaching and grasping in the direction of the object.[21] These cells are called "mirror cells," or "mirror neurons," because they fire the same way whether the action is being performed by the individual or whether they are simply observing the behaviour in another. This mirror system could underlie many of our social learning capabilities, and speech itself is commonly thought of as part of this mirror system.[22] Fortunately, even if manual gestures are a behavioural fossil, they don't appear to be doing us any harm. At worst, we might waste a little energy gesturing wildly while speaking on the telephone over Bluetooth in a vehicle to a listener who cannot see us.

TAKEAWAYS

Gesturing while speaking is natural, unconscious, and reveals the functional differences between the two halves of our lopsided brain. Because of the left hemisphere's dominance of speech *and* its control of the right hand, we tend to gesture with the right hand while speaking and often "listen" with the left hand. Some of this gesturing during speech could be from "spreading activation" as the part of the brain that controls the hand adjoins the part that controls the mouth. It also appears that language first developed through gestures, later evolving into oral/aural speech. The manual gestures we use today to accompany the spoken word could be a behavioural fossil from prehistoric communication.

10

Turning Biases: Things That Go Bump on the Right

Do I turn left, when nothing is right? Or do I turn right, when there's nothing left?

— ANONYMOUS

nspired by studying rotating systems such as waterwheels, French mathematician Gustave Coriolis[1] first described the natural forces that govern rotational systems in 1835.[2] Although his work was mostly focused on the movements of huge masses of water or air on a global scale, most of us refer to the effect bearing his name after observing the swirling of a very small body of water. The Coriolis effect is clearly evident in the direction of water after we flush a toilet. Coriolis is often credited with having both the first and last word about the impact of the Earth's rotation on other rotating systems, but much of the groundwork was laid two centuries earlier.

Before we could develop models of how the Earth's rotation could influence the motion of terrestrial bodies, including our own bodies, we first had to establish that our planet is round. It pains me to acknowledge that some supposedly modern-day humans have forgotten this scientific fact and have apparently never witnessed a lunar eclipse, watched a ship disappear while sailing out to sea, changed time zones, or witnessed other first-hand evidence of the Earth's curvature. Despite the misguided claims of Flat Earthers, we know that our planet rotates on its axis from west to east so that the sun, stars, and every other celestial object appear to move from east to west across the sky. Viewed from the North Pole, the world rotates counter-clockwise, but seen from the South Pole, the rotation is clockwise. If you enjoyed *The Daily Show* with Jon Stewart as much as I did, you'll have noticed the opening credits with the world spinning the wrong way many times. When the show changed hosts, the introductory graphic was corrected.

It might seem odd to start a chapter on human turning behaviour with a discussion of the Earth's rotation, but the Coriolis effect appears to influence human movements, at least in an experimental model.[3] Our brief survey of turning preferences in humans will consider the collective behaviour of groups of people as well as turning by individuals. We will look at turning behaviour from the ancient to modern worlds and across the human lifespan from unborn child to senior citizen.

The tendency of humans to turn our heads to the right is one of our earliest lopsided behaviours.[4] It is clearly visible after 38 weeks of fetus gestation, long before culture or social learning can influence an air-breathing child outside the womb. This turning bias persists throughout our lives. If we ask an average adult to walk down an empty hallway, turn around, and return, chances are he or she will rotate to the right. We can see evidence of this rightward bias when we drive, enter a store, play sports, or even dance. Most ancient dances have circular motions that tend to move in a clockwise (rightward) direction. Our tendency to turn right is even referenced in popular culture, such as Derek Zoolander's famous (and fictional) inability to turn left on the fashion runway in the movie named after him.[5] Why do we prefer to turn right?

Let's start by differentiating between turning, rotating, and circling. When I refer to rotation in this chapter, I am describing the partial or full spinning movement around the central axis of one's body.[6] Turning is different; it is the act of deviating from a straight line while moving and starting along an alternate path. Circling results from a series of turns that collectively form a complete circle around an external (outside one's body) reference point. All three forms of movement have some common themes in our lefts and rights, but let's start with circling.

Like most other ancient cultures, early Greek and Egyptian dances were circular, and most archaeological accounts of those celebrations describe the movement in a clockwise (rightward) direction.[7] European dances around the maypole and Breton dances also exhibit this same rightward circling.[8,9]

Circling in non-human animals has also been studied extensively, especially with regard to animal models of neurological diseases and drug addiction. For example, if we give an animal drugs that increase dopamine levels (common drugs of abuse that do this include cocaine and methamphetamine), the animals will tend to circle to the left.[10]

One of the earlier reports of human turning biases was by A.A. Schaeffer in 1928,[11] who blindfolded people and reported "spiral movement" during unsighted walking. In this brave experiment, people had to attempt to walk, run, swim, row, or even drive an automobile in a straight line while blindfolded. Schaeffer noted that individuals were often consistent in their turning directions across tasks but didn't document any population-level systematic lateral biases in turning behaviours.

Five years later, American psychologist Edward Robinson[12] published a report detailing turning asymmetries among American museum patrons. Across different museums in various cities, he noted that 75 percent of the visitors bore to the right upon entry to the building, even though most of the institutions were planned with the entrance displays on the left and patrons were often faced with signs directing them in the opposite direction. In those cases, the patrons "continued to walk first to the right and then around to the left."[13] This report from Robinson provides the reader with very few details about his methodology, but he does share several explanations for this curious contradiction. One potential explanation is that museum planners favour blueprints to plan exhibits, which tend to move from the left to the right side of the page. When the plans are translated into real-world spaces, they can produce exhibits contrary to the natural flow of patrons.

However, turning biases are present long before we start visiting museums. In a study of 72 10-week fetuses, Peter G. Hepper and colleagues[14] found evidence of very early lateralized (right-favouring) limb movements. By 38 weeks of gestation, the rightward-turning preference is already established, well before the baby is even born.[15] This makes turning biases one of the earliest lopsided behaviours to develop,[16] at least among those few observable in the womb. The turning bias also appears to be influenced by the length of time in the womb. The normal rightward turning preference is less prevalent in babies born preterm (less than 30 weeks of gestation).[17] For full-term babies, there's even some evidence that we can predict later hand preference based on earlier head-turning partiality.[18]

In the days immediately following birth, infants tend to lie in a supine position with the head turned to the right.[19] This position is preferred at rest but also if the infant is stimulated.[20] The bias is readily observable even two days after birth, and the infants who turn their heads to the right spend 70 to 80 percent of their time in this position. As a result, these right-facing infants also have much more visual experience (and sensory-motor feedback between eye and hand) with their right hands compared to the lefts, which has obvious implications for the development of handedness, particularly right-handedness.[21]

H. Stefan Bracha and colleagues[22] conducted a series of experiments about human turning biases using a "human rotometer," an automatic device they developed for measuring rotational behaviour as people go about their normal days.[23] This rechargeable device was worn in a belt-mounted calculator case, which they calibrated to magnetic north using a compass. Research like this was a lot more complicated and expensive before everyone carried their own gyro-servo, GPS-capable personal computers (i.e., cellphones) at all times! Similarly to the original report by A.A. Schaeffer,[24] Bracha found that individual men and women tend to consistently turn leftward or rightward. He also reported that men rotated more to the right than women.

Other researchers have also detected sex differences in turning, but these results aren't consistent in direction. Although some have found that both men and women prefer to rotate to the right, others have reported that right-sided men (right-handed, right-footed) tend to rotate to the right, whereas right-dominant women turn to the left. Why the different findings between the studies? Perhaps the turning bias in women is influenced by the phase of the menstrual cycle.

To investigate this possibility, Larissa Mead and Elizabeth Hampson compared the turning biases in women during the midluteal and menstrual phases of the menstrual cycle.[25] They collected saliva samples from 48 female students at the University of Western Ontario who weren't taking oral contraceptives (because those pills change the concentrations of sex hormones) and used a radioimmunoassay to determine levels of estradiol and progesterone to identify the phase of menstruation in each participant. Overall, the women tended to prefer turning to the right, but the women at the midluteal phase of the menstrual cycle demonstrated the weakest turning biases. Therefore, whatever the mechanism is behind our turning biases, it looks as if they're subject to modulation by ovarian hormones.

Schaeffer, in his 1928 study of turning biases,[26] tested blindfolded people while they walked, ran, drove, rowed, and swam. An American study examined turning biases during a virtual swimming task modelled after a common test of maze learning in rodents. In Richard Morris's water maze[27] (see Fig. 52), small rodents such as rats or gerbils were placed in a pool of murky, usually milky water; their task was to find the submerged platform just under the liquid's surface. Because the platform was hidden, the first time they found it was purely trial and error. They tended to swim around erratically, trying to find a way out of the "maze" until they bumped into the platform and could rest on it, thereby solving the puzzle. Subsequent trials were typically much easier for the rodents. When they were placed in the water, they tended to use the cues around the room to orient themselves and swim directly toward the location where they'd found the platform previously.

There are several "human" variants of this task, even one involving a large opaque pool! However, a more common way of testing this kind of wayfinding with humans is to use a water maze in a virtual-reality computer simulation. Nobody drowns or even gets wet. Take another look at Fig. 52. See all those crazy-looking turns on the first trial? What a great opportunity

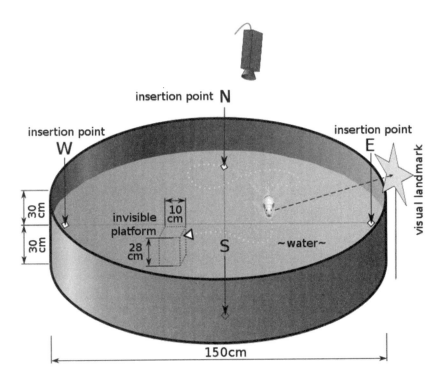

Fig. 52: An example of Richard Morris's water maze task. On the right side of the pool, a rat searches for the hidden platform and eventually finds it on the left. On subsequent trials, the animal uses cues around the room to guide it directly to the location where it previously found the platform.

to study turning biases! In Detroit, Michigan, in 2014, Peng Yuan and colleagues[28] did just that. They had 140 right-handed adults (18 to 77 years old) "swim" the virtual Morris water maze and then compared those results to scans of their brains to look for relationships between the relative size of parts of the brain and performance on the virtual maze. Men showed more of a leftward turning preference, whereas women usually exhibited rightward turning biases. People with larger movement-related brain areas (such as the putamen and cerebellum) on the right tended to turn to the right, whereas those with leftward biases in the brain also usually turned left.

Our turning biases are also influenced by factors such as the speed of movement, handedness, and training/practice with turning. When walking relatively slowly, students on a T-shaped runway were equally likely to turn left or right while returning. However, when running on the same runway, the people moving at high speeds usually veered left.[29] A study in Belgium produced similar findings.[30] When 107 adolescents were observed walking and running back and forth between two lines 9.5 metres apart, the group exhibited a general preference for leftward turning, and the bias was greatest during running (71 percent leftward) as compared with walking (59 percent).

Handedness matters, too. During spontaneous walking, 41 adults were monitored with a rotometer.[31] Most right-handers demonstrated a significant leftward bias, whereas left-handers didn't exhibit any bias whatsoever. In an Australian study of male left-handers and right-handers and female left-handers and right-handers, John and Judy Bradshaw[32] had their participants wear blindfolds and earmuffs while performing rotating and turning tasks. The right-handers had a rightward turning bias, but the left-handers tended to turn left. Similarly, when rotating, right-handers usually rotated more to the right, whereas the bias was reversed in left-handers. However, when asked to walk a straight line, all four groups usually deviated to the right.

Classical dance training also appears to modulate our turning biases. In a study of classically trained versus novice dancers,[33] untrained girls tended to show a leftward bias (58 percent) in turning. The vast majority of the trained dancers preferred turning to the right (only one turned to the left), suggesting that dance training, and the prevalence of clockwise dances, might modulate the effect.

Although most of us are functionally asymmetrical, we tend to be quite physically symmetrical. Not so with limb amputees, whose pronounced functional and physical asymmetries are the norm. In a study of turning biases in amputees, M.J.D. Taylor and colleagues[34] had 100 able-bodied and 30 trans-tibial amputees walk toward a mark 12 metres from the start point, turn around, and walk back. The able-bodied participants tended to turn left (opposite the most common hand and foot preference), whereas the amputees didn't exhibit any lateral preference as a group. This suggests that biomechanical biases can influence our turning biases.

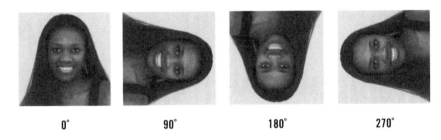

0°   90°   180°   270°

Fig. 53: Sample configuration of the head-rotation task employed by Sarah B. Wallwork and colleagues.

In addition to our tendency to turn to the right, we are also more accurate at perceiving rightward body rotations than we are at discerning otherwise identical movements to the left. In 2013, Sarah B. Wallwork and colleagues[35] presented 40 photographs of models with their heads turned to the left or right to 1,361 participants and asked them to identify the direction of rotation (it's harder than it sounds; see Fig. 53 for sample images). Rightward rotations were identified more quickly and more accurately, suggesting that rightward turning motor imagery is more accessible.

However, the rightward turning bias might not apply to everyone. Recall from several previous chapters that lateral biases can be influenced by native reading direction. Those of us who learned English as a first language are accustomed to visually scanning left to right, but individuals who learned scripts right to left (Arabic, Urdu, Hebrew) scan likewise. In Turkey (Ottoman Turkish is also read right to left), Emel Güneş and Erhan Nalçaci[36] assessed turning biases in 31 children between 7 and 13. They used the same rotometer-type gear I described earlier. The rightward turning biases noted among other cultures were gone entirely. Most of the children preferred turning left instead, and this bias was stronger among boys than girls.

But it isn't just Turkish children who prefer turning to the left. Observing a daycare setting, H.D. and Kaaren Day[37] studied rotational play among 67 Texan children aged three to five years old. They employed a couple of tracks marked out with masking tape where the children could walk, run, or use a tricycle to circle the track. Regardless of the method children chose to circle the track, they tended to prefer a counter-clockwise (leftward turning) path, contrary to the right turning biases we typically see in adulthood. (But not always — some studies of adults find leftward turning biases, especially among left-handers or otherwise unusual right hemispheres.[38–41]) Many organized sports involve counter-clockwise motion, including Olympic events such as running, speed skating, and cycling. Even baseball integrates counter-clockwise motion. We discuss these biases in more detail in Chapter 12.

So where do these turning biases come from? The ones in humans and other mammals are typically attributed to asymmetries in levels of the neurotransmitter dopamine (DA) in brain structures, such as the striatum, responsible for initiative movement. We know that in many species DA levels are higher in the left striatum, and because the left hemisphere controls the right side of the body, this results in right-sided turning. Post-mortem examinations of human brains have also found an asymmetry in DA levels, with greater DA in the left globus pallidus (one of the structures within the striatum that's responsible for initiating movements such as walking).

This simple, movement-focused explanation is parsimonious and intuitively appealing. Unfortunately, our rightward turning biases are probably more complicated than that. We've already discovered that the effect seems to depend on many factors, including age, sex (and possibly sex hormones), handedness, and maybe even native reading direction. However, what if this rightward-turning effect emerges even when we take asymmetries in the movement system out of the equation?

Before we do that, let's first consider the results of a survey conducted by Oliver Turnbull and Peter McGeorge in 1988.[42] They asked 383 participants to recall whether they'd recently bumped into anything, and if so, which side of their bodies was involved in the collision? The people surveyed showed a small tendency to recall right-sided collisions. Furthermore, the same people completed a clinical test called "line bisection" in which they were presented with horizontal lines and asked to indicate their midpoints. Neurologically typical people tend to be quite accurate at this task, choosing points very near the middle, but when they miss the middle, they usually do so to the left of centre (overestimating the length of the rightward portion of the line). This overestimation is often called pseudoneglect[43] because it resembles the clinical condition of neglect, when an individual suffers a brain injury and then fails to be aware of one side of space (usually the left side). Pseudoneglect is much more subtle than clinical neglect; instead of being characterized by inattention to the left side of space, it results from allocating too much attention to the left side.

People with stronger pseudoneglect were *more* likely to report right-side collisions, which initially seems contradictory. However, if we tend to pay more attention to objects on the left side, this could cause us to partially ignore objects on the right, resulting in increased rightward collisions through the pseudoneglect of that side of space. As we will see in Chapter 12, which is about sports, this pseudoneglect effect is clearly visible across all kinds of scenarios.

But I promised to take the movement system out of the equation while still looking at these turning biases. The study mentioned above might not have required *real* collisions, but people still imagined movement. How then can we get people to move without moving? Well, we do it the same way we help people who can't move to move — with a wheelchair. In a series of clever studies, Australian Michael Nicholls and colleagues studied lateral bumping requiring different types of movements. In the first study,[44] he had almost 300 university students walk through a narrow doorway and recorded the side of collisions with the frame. The most frequent outcome was that the student succeeded in not hitting the doorway. This happened 38 percent of the time. Other students weren't so lucky. Some collided with both sides of the doorway, a result that occurred 13.5 percent of the time. The study's focus was on the one-sided collisions, though. Right-side collisions were much more frequent (29.6 percent of the time) than left-sided ones (18.7 percent).

But this first study had people walking through the door. A follow-up[45] employed the same type of task, but the participants had to pilot an electric wheelchair (using handlebars) through the narrow gap instead of walking through it. It still took *some* movement to control the chair, of course, but the task was mostly a visual-perceptual one now instead of a movement task. What happened? People still missed to the right. In a final study, the same research group asked people to use a laser pointer to indicate where they thought the middle of the door gap was.[46] Participants still misjudged the centre of the doorway to the right, even when they were no longer required to pass through it. Collectively, these results indicate that the rightward bumping effect could be due to biases in how we *perceive* obstacles, like doorways, instead of how we *move* about them.

TAKEAWAYS

Under most circumstances, we prefer to turn to the right. This bias appears very early in life, perhaps even before birth. It influences how we behave when we enter a new space (museum, movie theatre, classroom), how we interact with one another (through social touches like hugs), and how groups of people coordinate movement together through dance. This turning bias is influenced by handedness, age, gender, and possibly even native reading direction. As noted in other chapters, turning biases might cause or contribute to other lateral biases such as for kissing, hugging, seat selection, and biases in sport. Although these turning biases clearly involve movement, there is obviously a perceptual side to the effect. So what can we do with this information? When Edward Robinson first reported turning biases among museum patrons in 1933, he suggested that we optimize the physical layout of public spaces, and the "objective standards of educational efficiency can replace the hunch of the artist, the poet, and the advertising man."[47] Ninety years later, we still have the same opportunity. We should be leveraging the results of these scientific studies to inform our design of public parks, museums, schools, and shopping centres. We can even leverage the impact of these biases by strategically placing items within these spaces. Knowing that people naturally turn right when entering a space, we can plan for traffic flow that accommodates this bias, and even place items, such as signs or products, to the right if we want them to be noticed first.

11

Seating Biases: 2B or Not 2B?

If you're offered a seat on a rocket ship, don't ask what seat. Just get on.

— SHERYL SANDBERG, COO of Meta
Platforms (formerly Facebook, Inc.)

Choosing a place to sit down should be easy, but it so often isn't. If we are attending a Broadway play or a playoff game, we probably want a good view. On the other hand, maybe we don't want to be seen, especially if we are "sick" from work during the afternoon baseball game! In a fourth-period class, there might be a cute girl or boy to sit close to but not conspicuously near. Perhaps that annoying person from yoga is also going to the same movie, and we want to avoid him or her. Maybe we want to sit where there is more legroom, or perhaps close to the exit or window, or near the heater/air conditioner. We might need a place to plug in our phones or computers, leading us to scan the room for an outlet. If we are on an airplane, we might want the extra legroom provided by the emergency exit row, but we might not be up for the responsibility of pulling that big handle if things go sideways! These are just some of the factors to weigh when searching for a seat when we are alone. Choosing seats collaboratively with someone else adds another layer of complications.

Throughout this chapter, I will consider the left-right biases in seating behaviour across a wide variety of circumstances. Simply looking at seating decisions in terms of right/left sides might seem like a gross oversimplification of a multifactorial process, and that's because it is. However, we will discover there are some consistent lateral biases in how and where people take their seats. These discoveries have been made using a wide variety of research methods, but the studies generally fall into two types. Some have actually observed real people taking real seats in the real world (naturalistic observation), whereas others have required people to imagine taking a seat or to select a fictitious seat on a seating map of an airplane, theatre, or stadium.

We will also learn that seating biases depend on the type of venue. Most of the early research in this area focused on the elementary school classroom, relating seating position to academic performance. If we have long suspected that keeners sit in the front, we will be pleased to find those feelings backed up by hard evidence. Later research has concentrated more on

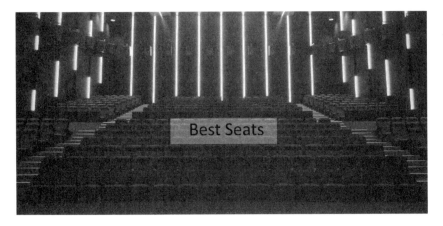

Fig. 54: The ideal seats in a movie theatre, optimizing the aural and visual experience.

decision-making in movie theatres and large commercial aircraft, with the former often happening "in the wild" and the latter usually taking place when confronted with a seating chart, either on paper or online. However, in an effort to be more commercially viable in the Netflix and post-Covid-19 age, movie theatres have moved to an assigned-seating (and premium-seating) business model that often utilizes online seating selection. Further complicating matters, the distinction between aircraft and movie theatre seating used to be blurry. On my first commercial flights as a child, I remember being quite excited about the prospect of an inflight movie (even more dated than that experience, there were also smoking sections on aircraft at that time). These days, most commercial aircraft have video screens on board, but the proliferation of smaller screens and individualized inflight entertainment has largely done away with the large-screen inflight movie. This is progress.

Of course, we can only choose a seat if we actually have a choice. When a flight is 99 percent full, taking that seat in the middle of a row, adjacent to the bathroom, or at the very back of the plane might be the only way to get home for Thanksgiving. There are also cultural norms and written (or even unwritten) rules about who must sit where at a wedding, funeral, political event, or Sunday church service.

If we google how to choose a good seat in a movie theatre or aircraft, we tend to get pretty complicated advice. For movie theatres, we find advice to sit centrally, far enough back to avoid neck strain and motion sickness, but close enough to see all the details. Taking into account the sound mixing optimized for the middle seats (both front/back and left/right), this results in a recommendation to sit centrally and two-thirds back (see Fig. 54).

Advice for seating in airplanes becomes even more complicated. Other than the general agreement that most people like to sit close to the front (presumably to be off first), the other factors depend on individual preferences. Do we want legroom (aisle seats) or a view (window)? Do we want to be near or far from the bathrooms? If there are screens in the cabin, do we want a good view or do we wish not to be bothered with the inflight presentation? Living in a post-Covid-19 world, some choose seats with the best ventilation and minimal contact with other passengers and airline staff, leading to a preference for window seats near the front of the aircraft.

Many passengers favour bulkhead seats located directly behind physical barriers in the plane (walls, screens, curtains) because there are no passengers directly in front of them and nobody can recline into their laps as they savour the presumably excellent inflight cuisine. One of the few areas of consensus concerns the middle seat in any aisle. Almost everybody agrees that's the *worst* place to be, even if it entitles us to two armrests (although this "rule" isn't universally known or followed). We should also be wary of seats just ahead of bulkheads, since they often cannot recline very far, if at all.

CLASSROOMS

Most of the earlier studies on seating biases focused on classroom settings, investigating links between seating biases (left/right, front/back) to academic performance. For example, in 1933 Paul Farnsworth[1] showed students a seating chart and asked them to identify where they preferred to sit in classes, with four different instructors across three different academic subjects. Comparing the academic performance of the students according to seating position, he found that the students with the best grades tended to sit toward the front of the room and slightly to the right of centre. Instead of concentrating on the perceptions and preferences of the student, Farnsworth's explanation for this effect focused on the instructor. He reasoned they would pay more attention to the students at the front of the class. Furthermore, the position of a right-handed instructor in front of a chalkboard (smartboards and LCD projectors were relatively rare in 1933) would tend to be toward the right side of the room, so sitting to the right would put the student even closer to the instructor. Since the original study, several research groups have replicated the "keeners-sit-in-the-front" effect.[2,3]

Later studies focused more attention on the students themselves, along with the teaching material. In the early 1970s, some researchers believed they could infer the activation of one hemisphere of the brain by looking at the direction of a person's gaze while thinking.[4] Looking to the left was supposedly indicative of right-hemisphere activation and vice versa. Following this logic, Raquel Gur and colleagues[5] examined lateral eye movements in 74 students and compared those biases to where the students preferred to sit. They measured eye-movement directionality by asking the students either verbal or spatial questions, recording which direction the students glanced while answering. They thought seating biases were "symptomatic of easier triggering of activity in the hemisphere contralateral to the direction of eye movement."[6] In other words, rightward eye movers should prefer sitting on the left, whereas leftward ones should prefer sitting on the right. Sure enough, the "left-eye movers" (people exhibiting 70 percent or more of eye movements to the left) preferred to sit on the right side of the classroom. Conversely, the "right-eye movers" chose to sit on the left side of the classroom. Therefore, students tended to seat themselves in positions that facilitated their habitual modes of information processing.

The follow-up study in 1976 asked a more eye-catching question when the same researchers investigated the relationship between psychopathology and classroom seating preferences.[7] This might seem like a really unusual link to make, but even prior to the 1970s many studies had already linked many mental disorders, especially emotional ones, to dysfunction or injury in the right hemisphere of the brain.[8] The researchers gave a 124-item questionnaire about mental illness, spanning 65 different psychiatric disorders, to a couple of hundred undergraduate students in an introductory psychology class and compared the scores for the people seated on the left to those on the right. The males sitting on the right side of the classroom scored higher on psychopathology than those on the left, but the researchers found the opposite effect in females. The females on the left side of the room had higher psychopathology scores than those on the right. Therefore, right hemisphericity was associated with greater psychopathology in males.

We can also investigate the relationship between seating biases and personality or learning style by identifying people with a left or right seating bias, then consider what is different between the two groups. That is exactly what Larry Morton and John Kershner did in 1987 at the University of Toronto,[9] predicting that children would develop seating strategies (and the corresponding systematic biases in seating) that maximize learning and minimize the effort required to learn. Instead of asking students to indicate a seating preference under a certain condition, Morton and Kershner identified those who exhibited right or left seating biases, then investigated what was different between these two groups. They predicted that children would develop seating strategies (and systematic biases where they sat) to maximize their learning.

Because the right hemisphere is normally specialized for emotional and spatial content versus the left hemisphere being typically specialized for language processing, Morton and Kershner anticipated that children with a more language-based learning style would want to expose the left hemisphere (right side of space) while learning, whereas children with a more visual-spatial learning style would want to preferentially display the right hemisphere (left side of space). The researchers also projected that children sitting on the right would be better spellers and make more phonetically inaccurate errors, whereas those on the left would make more phonetically accurate errors.

The teachers administered spelling tests while noting the side of the classroom and gender of each child. They analyzed errors for phonetically accurate misspellings (implying relatively greater reliance on phonological rather than on visual processes) and phonetically inaccurate misspellings (implying relatively greater reliance on visual rather than on phonological processes).

The prediction that children sitting on the right were better spellers turned out to be true. Those rightward-sitting children also relied on non-phonological processes rather than phonological processes to generate spelling. In contrast, the leftward-sitting children scored more poorly on spelling, but only the female kids on the left demonstrated a decrease in the use of non-phonological processes. The Morton and Kershner concluded that rightward-sitting children might depend more on right-hemisphere processing when learning, which augments visual memory for whole words.

My own research group has also studied seating biases in classroom settings, but instead of focusing on elementary school kids and spelling ability, we studied college-level students.[10] Because these classes tend to be lecture-based and require students to think analytically, we expected this content to selectively draw on the language resources of the left hemisphere of the brain. Remember that content presented to the observer's right side is predominately processed by the left hemisphere. Therefore, we assumed that our college students would prefer seats on the left side of the classroom so that more of the content would be presented from the right side of the observer, preferentially exposing the left hemisphere.

To assess the seating biases of our college students, we identified bilaterally symmetrical classrooms where the locations of entrances and exits or amenities weren't likely to drive seating patterns on their own. Over a nine-week period, we visited these classrooms at the beginning of each class. If the capacity of the room was 50 percent or greater (allowing considerable choice of seats) when class started, we took a photograph of the class from behind and assessed the seating pattern. This resulted in samples from 29 different classrooms across 41 occasions, and when we coded and analyzed the data, we discovered a leftward seating bias among our college students (see Fig. 55).[11]

This was a unique and encouraging result for us. Unlike the earlier studies of elementary school children that reported a bias favouring the right side of the room, our study of college students discovered the opposite effect, but

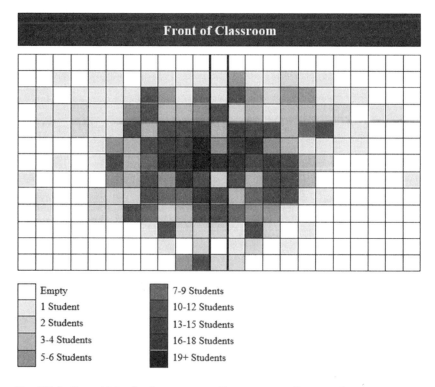

Fig. 55: Leftward bias in classroom seating among college students.

in a very different circumstance. Recall from the earlier studies of classroom seating that students tended to sit on the right and that those on the right usually scored better on spelling tests. Unfortunately, we couldn't compare the seating positions of our students to their academic performances in the classroom, so we cannot say whether the leftward bias we observed was also linked to improved performance in the class. Having taught college-level courses for more than 20 years, I can tell you that student spelling isn't getting any better.

MOVIE THEATRES

But let's get out of the classroom and head to the movies instead. Movie and lecture theatres might look physically similar, but the content found at the front of the room is, hopefully, quite different. In 2000, Bulgarian anthropologist George Karev published a study[12] concerning seating biases in movie theatres, but in addition to asking where people preferred to sit, he also wanted to examine the influence of handedness. (As an aside, almost every time I describe a real-world lateral bias to people, they ask whether it is reversed for left-handers. The answer is usually no.)

Karev conducted a rather ambitious study using five seating maps of cinema halls and had hundreds of students (264 right-handed, 246 mixed-handed, and 360 left-handed) indicate on the maps where they preferred to sit. Regardless of handedness, all three groups favoured rightward positions when selecting theatre seats, which suggested a perceptual bias. This bias was strongest in right-handers (88.26 percent choosing the right), weaker in mixed handers (66.67 percent picking the right), and weaker still among left-handers (57.50 percent selecting the right). Karev concluded that the right-seat preference presented itself because a rightward position facilitates leftward orientation of attention and thus fosters activation of and exposure to the emotionally dominant right hemisphere. He also used the term *expectation bias* to describe the seating positions because individuals decide on seats based on preparation for forthcoming content.

In a 2006 follow-up study, Peter Weyers and colleagues[13] used a procedure similar to Karev's and also found a right-side bias for seating preference. However, the effect disappeared when the theatre was presented in a non-canonical perspective. If the portrayal of the screen or stage on the seating map was moved from the top of the page to the side or bottom, the right-side seating bias disappeared. Instead of simply finding a right seating bias, Weyers and colleagues argued that the "real" bias was toward the right side of the paper, not the right side of the theatre. If the movie screen was no longer presented at the top of the paper, their participants chose the left side of the theatre but the right side of the paper. The researchers also argued against the expectancy hypothesis, claiming that the right seating bias was just another manifestation of a general bias toward the right side of space.

Have you ever gone to a movie you were not looking forward to seeing? I can certainly think of stories that I felt a moral obligation to see (*Schindler's List* comes to mind), and even if the movie is made exceptionally well, going to it isn't necessarily a pleasant experience. I could also list several kids' films I have suffered through but naming them here would surely get me in trouble with my own children. In Japan, Matia Okubo[14] followed up on the Weyers study but wondered whether the seating biases would be different if people were positively motivated to see the movie. When he tested people who actually desired to see a film (positive motivation), he found the same right-side bias for theatre seating maps. However, when the positive motivation to see the film disappeared, so, too, did the rightward bias! This seems to indicate that people have to want to engage their emotionally dominant right hemisphere when viewing a movie before the rightward seating bias emerges.

My own research group has also studied seating biases in movie theatres,[15] but our study differs from earlier work in an important way. The earlier theatre-seating studies used paper seating charts to determine seat choice, and assumed that the choices people make on paper are the same as the ones they would pick if they walked into a crowded theatre. In our study, we actually observed real people going to a real movie in a real theatre and recorded where they really sat. We used a method similar to the one in our classroom study described earlier in the chapter. For movies that were less than 50 percent full (there were still many seats to choose from when the film started), we took a picture of the

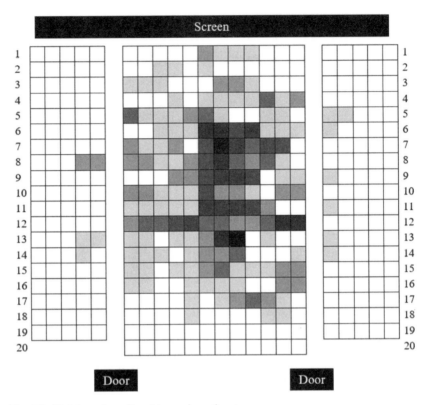

Fig. 56: Rightward seating biases in a theatre.

seating position at the start. Just as in the earlier studies using seating charts, our actual moviegoers preferred to sit on the right side of the theatre (see Fig. 56).

Recall that the right hemisphere of the brain is dominant for emotional processing, especially when sorting negative emotional content. It appears as though moviegoers like to preferentially expose their right hemispheres when going to a film, sitting on the right side of the theatre, with most of the screen to their left sides (which are preferentially perceived by the right hemisphere). People *expect* emotional content when they go to movies, and this expectancy influences their seat choices.

AIRPLANES

Choosing a place to sit for a quick lunch at a mall is not a big deal. Even if we opt for a drafty spot, or perhaps one on the path to the washroom, the pain will be short-lived. However, selecting the wrong spot on a transoceanic flight can be a lot more painful. It's a decision we can regret for many hours and remember for years. Unfortunately, the research on airplane seating is somewhat mixed, and it is difficult to give concrete advice on where to sit based on the few available studies. This is especially surprising because airplanes are asymmetrical by design. Every commercial airliner I have seen has the main entrance/exit on the left side of the plane. Another factor complicating the study of seating preferences on planes is the location of amenities such as video screens. Airplanes used to be flying movie theatres, but the prevalence of large, shared screens has mostly gone the way of Blockbuster Video.

A couple of commercial airlines have actually released studies of their own on seating preferences, and even these have resulted in conflicting information. When easyJet launched a new online booking system in 2012, it issued a press release indicating that passengers preferred to select leftward seats using the new system. However, two years later, easyJet released the cleverly titled report "2B or not 2B," in which it indicated that passengers favoured bookings on the right. A rightward seating bias on wide-body airplanes was also reported by British Airways passengers in a survey.[16]

A large-scale Australian study of aircraft seating preferences was conducted by Michael Nicholls and colleagues in 2013.[17] They analyzed real seating patterns for more than 8,000 seats across 100 airplanes and found a seating preference for the left side of the aircraft. They concluded that this leftward seating bias was likely the result of a rightward turning bias, because when turning right after entering the aircraft near the cockpit (entering from the left side) and proceeding toward the back of the plane, a right turn will result in a leftward seating position.

Real-world studies of how real people behave under real conditions have lots of advantages, such as validity, but they also have some major disadvantages. Most notably they result in the loss of control over some of the factors that could influence the behaviour of interest. The discrepancies in past

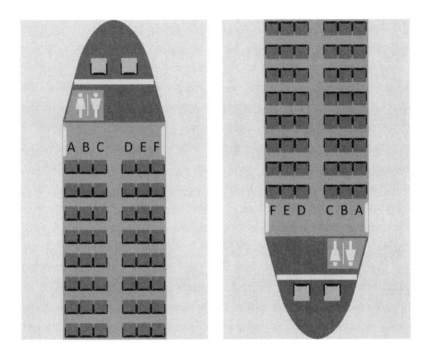

Fig. 57: Using this seating map for a fictitious aircraft cabin, with the front depicted on top and A-F indicating the seat choices from left to right, people indicated where they would prefer to sit. Regardless of how the map was oriented, people preferred seats at the front of the plane, on the right side.

studies could have been related to the positioning of screens in the cabin. There are also language cues that could bias travellers toward choosing seats on the left. In narrow-body aircraft, seats are identified using letters, with the leftward three seats labelled A, B, and C, and the rightward seats marked D, E, and F. The passengers might simply be preferring the lower letters, or the letters read first, resulting in a leftward seating bias. It is also not completely clear how airline companies "release" or allocate seats to customers and which algorithms are at play for real-world seat selection. In this respect, the studies using fictitious seating maps for imaginary planes and fantasy trips have the advantage. All those variables can be controlled and manipulated.

In an attempt to do just that, a recent follow-up study in the United Kingdom deployed its own seating maps for fictitious flights and presented them in different orientations.[18] In some cases, the right side of the screen represented the right side of the plane. In other cases, the right side of the screen represented the left side of the plane (see Fig. 57). Regardless of the orientation of the seating map, the prospective passengers on these fictitious flights preferred seats toward the front, at the window or aisles (not in the middle), and on the right side of the plane. In other words, they found the exact opposite of that reported by the Australian group. Why? Is it because the Australians were upside down and left is also right? Just kidding ... sort of.

TAKEAWAYS

We surveyed seating preferences in classrooms, movie theatres, and airplanes and found some conflicting information but also some themes. In an elementary school classroom, keeners tend to sit near the front and slightly to the right. In college classes, this left-right trend appears to reverse direction, with most students choosing seats on the left. In movie theatres, patrons tend to choose seats on the right side of the theatre, especially if they really want to see the movie. At least some of this effect might be due to rightward biases on the seating maps instead of real-world seating behaviour. Finally, the research on seat choice on airplanes presents us with mixed results. Some studies find rightward seating biases regardless of the orientation of the seating map, whereas others have found left-sided biases among real-world Australian travellers. Of course, the act of choosing a seat varies tremendously across these contexts. When we walk into a classroom, seat choice is typically made in real time and in real space after quickly scanning the room. Picking a seat on an airplane is typically very different. That choice is usually made through a map on a screen, sometimes months before actually taking the seat, and that selection can be influenced by the orientation and labelling of the map, not simply the physical location of the seat. Which seat is best? It depends on the circumstances and how the seat is actually chosen — 2B or not 2B, that is indeed the question.

12

Sports: Competing the Right Way

I would choose Messi's left foot, Neymar's right foot, Cristiano's mentality, and Buffon's class.
— KYLIAN MBAPPÉ on the perfect soccer player

t's hard to imagine a better arena than the crucible of sport for studying the lefts and rights of human behaviour. Earlier in this book, I described the painstaking and creative ways that researchers have observed, coded, and analyzed how people hug and kiss one another in airports, or how mothers cradle their newborn babies shortly after birth. Recording and evaluating these types of behaviours is actually quite unusual. As a result, the passages within these scientific papers that describe their methods can be a very strange read. At times, the papers even seem voyeuristic. Not so with sports. Where else can you find detailed records of the handedness of an individual along with other vital statistics (sex, height, weight, age, et cetera) paired with objective measures of proficiency at a wide variety of tasks? For example, take baseball statistics. Anyone with internet access can plunge into detailed databases started in the 19th century, listing handedness, batting averages (among a whole host of other performance indicators), and even how long the athlete lived. Sports statistics are big business, but they are also a treasure trove of useful data for laterality researchers like myself.

There is another great reason to study lateral preferences in sports. It might help us solve one of the biggest mysteries in laterality research. We know that 10 percent of the human population is left-handed and that this 10/90 percent split has been relatively stable for centuries.[1] But we don't know *why*. Sure, we know that handedness runs in families[2] and that it varies a fair bit across cultures.[3] We even have some important clues about how and when handedness develops in the very young human, but that doesn't explain the dominance of right-handedness across centuries or the persistence of a left-handedness minority over those same centuries.

In Chapter 1, I detailed all the negative associations with left-handedness, including higher rates of birth trauma, autoimmune disorders, et cetera, along with some of the more positive ones (being overrepresented among the intellectually gifted, professional artists, et cetera). However, none of these associations or mechanisms tells us how left-handedness can be an advantage among a majority (90 percent) of right-handers.

One of the leading theories is quite simple. Maybe left-handedness provided, and still provides, an advantage in fighting. These days, in most developed countries, hand-to-hand combat is a relatively rare event. Despite the impression we might get from watching the evening news or even a Twitter feed, modern society is getting less and less violent all the time. Steven Pinker's book *The Better Angels of Our Nature*[4] does an excellent job of articulating all the ways the world has become a safer and less violent place over recent decades and centuries, if not millennia. However, our collective history as humans is a disturbingly violent one, and echoes of that history are readily observable in the few hunter-gatherer tribes left on this planet, such as the Yanomami in the Amazon rainforest.[5] Even a cursory read of the anthropological record of these hunter-gatherers reveals a dark lesson.

Violence is a common occurrence in these societies, and people (mostly men) who excel at violence are richly rewarded, both through material goods and success in passing along their genes to the next generations. Men from these tribes who have killed a number of other men tend to have more children.[6] Therefore, proficiency at violence doesn't simply keep these men alive; it also helps ensure their genes are well represented in the next generation.

According to the "fighting hypothesis,"[7-8] left-handers benefit from an advantage in confrontations of physical aggression, which balances out any adverse effects that might also accompany left-handedness, such as higher rates of autoimmune disease. The advantage in combat is primarily due to the reduced practice/familiarity for the right-handed majority in facing left-handed opponents. Right-handers could be less able to perceive aggressive movements from left-handers, less able to predict the outcome of a left-handed move, and even less able to formulate successful strategies versus a left-handed opponent. In a contest against a left-hander, the right-hander might even need to use mental rotation[9] to intentionally reverse a series of offensive or defensive moves.

This intriguing theory makes a number of obvious and easily tested predictions. Left-handedness should be more prevalent in males than in females because the "selection pressure" on males was, and still is, stronger for acts of physical aggression. We should also see more left-handers in fighting sports, such as boxing, wrestling, and mixed martial arts (MMA). Furthermore, the left-handers we do find specializing in these fighting sports should be better fighters, especially when matched against right-handed opponents. So let's look at some of these predictions and see how they pan out.

Are left-handers overrepresented in fighting sports? It certainly appears so. Consider fencing. In the 1981 World Fencing Championship, 35 percent of the athletes in the foil competition were left-handed, compared to approximately 10 percent in the average population.[10] Furthermore, the left-handed fencers were more likely to advance to the later rounds of the competition. This effect was not specific to 1981. Between 1979 and 1993, 44.5 percent of fencing tournament champions were left-handed.[11] In the 1980 Summer Olympic Games, left-handed fencers took all eight top places in the competition. Among the many fascinating and accomplished

Fig. 58: Edoardo Mangiarotti was a natural right-hander, but his father had him train as a left-handed fencer. He became the most successful fencer the world has ever seen.

left-handed fencers I could profile, one stands out over all the rest. Edoardo Mangiarotti was born in 1919 as a natural right-hander, but his father (Giuseppe Mangiarotti, himself the 17-time national fencing champion in Italy) had his son change to using his left hand. Following the switch, Edoardo went on to win 39 gold, silver, and bronze medals in subsequent world and Olympic championships, a record that still holds for the sport.[12–14] Left-handed fencers clearly have an "edge."

But this advantage is not limited to fencing. In a large survey of more than 9,800 professional boxers and MMA fighters, left-handers were over-represented and tended to enjoy greater success in the ring.[15] Both male (17.3

percent left-handed) and female (12.6 percent left-handed) boxers showed elevated rates of left-handedness. Success in boxing can be measured in several interesting ways. At the highest levels, it might be determined by prize earnings or the number of Las Vegas Boulevard billboards displaying the fighter's face. However, at lower levels, boxing success can be measured with the number of wins, knockouts, rounds with better scores than the opponent, or even a metric called the BoxRec score.[16]

Male boxers have higher win percentages and BoxRec scores. Female boxers also showed higher BoxRec scores (a more comprehensive measure of fighting ability), but not higher win percentages compared to their right-handed counterparts.

In MMA, left-handed fighters are also overrepresented (18.7 percent left-handers in a combined sample of male and female fighters). Even though the left-handed MMA fighters are not heavier or taller than the right-handers, they have higher win percentages, consistent with the fighting hypothesis.[17]

Left-handers are also overrepresented among successful wrestlers. In a survey of handedness among wrestlers competing in world championships, left-handers were no more common than right-handers among all the competitors surveyed (44 out of 440 athletes, mirroring the 10 percent one would expect from the general population). However, the left-handed competitors had fewer losing rounds than the right-handers and tended to progress further in the competition. Among medal winners, left- and mixed-handed athletes received 34 percent of the gold medals, 35 percent of the silver medals, and 27 percent of the bronze medals,[18] far greater than the 10 percent one would expect overall based on chance alone.

Left-handers are also overrepresented in some traditional martial arts, including karate and taekwondo.[19] In these cases, the left-handedness advantage also seems to result in more tournament victories and medals, at least among female competitors. The left-hander advantage appears to persist even in sports with a really loose connection to actual combat. In archery, arrows flung from a left-hander's bow tend to yield higher scores.[20] Keep archery in mind, though, because we'll circle back to it later in this chapter, taking a different angle on it.

If I could end the chapter here, I could leave the reader with a very simple story supported by strong evidence: left-handedness has persisted across the centuries because it provides an advantage in combat that equals or outweighs the potential disadvantages that also accompany left-handedness. But the story isn't that simple. One alternative to the fighting hypothesis is the much more awkwardly named "negative frequency–dependent selection hypothesis."[21,22] Despite having a much longer and more complicated name than the fighting hypothesis, this theory is even simpler. In a nutshell, it states that left-handed people are more successful because they are less numerous (frequent). Right-handed fighters tend to have less experience against left-handed opponents. Both hypotheses account for the data from combat sports quite well, but the story doesn't end there, at least not for this newer hypothesis.

The consensus is growing that left-handers are overrepresented and disproportionately successful at combat sports, but lefties also thrive in other sports. A wide variety of non-combative sports also appear to favour the left-hander. In cricket, some of the most successful teams have close to 50 percent left-handed batsmen, and the overall bias in the 2003 World Cup was 24 percent left-handed batsmen.[23] Other "fast-ball" sports also appear to prefer the left-hander, including baseball, soccer, basketball, volleyball, Australian football, water polo, and tennis.[24–32] In water polo, left-handers were most common among players on the wings, with 24 percent of male players and 34 percent of women in the 2011, 2013, and 2015 world championships.[33] Furthermore, the left-handed players took more shots and scored more goals. In Australian football, left-footed penalty kicks are more successful than right-footed attempts.[34] Admittedly, I just referred to sports such as water polo and Australian football as "non-combative," but anyone with experience watching or participating in these sports knows that although combat might not be the main objective of the sport, it can still be a factor. So let's turn our attention to tennis, historically, and *sexistly*, referred to as a "gentleman's sport."

In a survey of professional male tennis players competing in Grand Slam tournaments between 1968 and 2011, left-handers comprised only 10.9 percent of all players in the first rounds of play (about what one would expect based on the general population), but 17.1 percent of the finalists were left-handed, including 21.2 percent of the champions.[35] The year-end

Fig. 59: Professional tennis player Rafael Nadal was reportedly right-handed as a child but was encouraged by his uncle to learn to play tennis left-handed.

rankings of tennis players between 1973 and 2011 demonstrated a similar pattern. Left-handed players were not overrepresented among all the players (9.6 percent), but they were much more common among the top 100 players (13.4 percent) and especially among the top 10 players (13.8 percent). The same effect is present but less pronounced among female professional players.

The left-hander advantage in tennis isn't restricted to professionals. In a survey of 3,793 amateur players,[36] left-handedness was even less common than one would expect in the general population (6.8 percent for male players, 4.4 percent for females), but the frequency of left-handedness rose sharply with performance. The left-handed players were much more likely to compete at a high level and win their matches.

There are many charismatic, successful, and interesting left-handed tennis players I could highlight before moving on to discuss other sports, but the Spanish player Rafael Nadal (see Fig. 59) deserves special mention. Similar to the story of fencer Edoardo Mangiarotti that I told earlier in this chapter, Nadal is a very successful tennis player who competes left-handed but is purportedly a "natural right-hander." As a child, Nadal performed most skilled tasks with his right hand, including writing and throwing. He trained initially as a two-handed player and was "encouraged" by his uncle, Toni Nadal, to develop a left-handed swing, especially on the backhand side. With 91 tennis titles in hand, including 21 Grand Slam titles, and no end in sight, Toni's coaching appears to have paid off.

Why are left-handed tennis players more successful? Their advantage could be a general one, such as the one articulated in the fighting hypothesis or the negative frequency–dependent hypothesis, but it could also be something more specific to the game of tennis. After analyzing 54 professional tennis matches, a German research group[37] found that right-handers were less likely to hit the ball to the opponent's backhand (usually the weaker stroke) when playing a left-handed opponent. The same research group also had 108 tennis players watch pre-recorded tennis rallies and asked them where they would place the ball in the opposing court, measuring their tactical strategies (regardless of their ability to hit the intended shot). Just as the researchers found with the professional players, these players were less likely to intend to place the ball to the left-handed opponent's backhand. So at least part of the success of left-handed tennis players can be attributed to right-handed opponents employing less-effective strategies against them. Of course, this same type of advantage might be found in other sports, combat and otherwise, and it could even be the mechanism for the negative frequency–dependent effect.

In soccer, coaches explicitly train young players to develop proficiency kicking with either foot. In fact, a strong one-foot bias in a player is often regarded as a result of poor coaching or insufficient practice. In European professional soccer, players who are proficient with both feet actually earn more money (salaries between 13.2 percent and 18.6 percent higher depending on the league).[38] However, specific tasks and plays in soccer create a demand for players proficient with kicking with the left foot. As we learned in Chapter 1, the prevalence of people with left-foot preferences is around 20 percent of the population.[39,40] An analysis of 19,295 kicks by 236 FIFA players during the 1998 World Cup found that 79.2 percent of these elite soccer players preferred the right foot, the same proportion found in the general population.[41] However, in soccer, the demand for left-footed players approaches 40 percent. Predictably, players' foot preference has an effect on their chances of playing at an elite level. In a Dutch study of national soccer team selections, a cohort of 280 young players (aged 16 at the beginning of the five-year study period) were followed through the U16 to U19 teams to measure their success in reaching the national youth team. Being left-footed clearly gave players an advantage to reach that club, since 31 percent of the national youth team was left-footed.[42]

Another sport requiring "bilateral competence" (manipulating the ball with hands instead of feet) is basketball. Players who are competent with either hand can defend or score with much greater proficiency than single-sided players, making them more adaptable to different situations and more difficult to defend against. Surprisingly, a large-scale study of National Basketball Association players (3,647 participating in at least five games between 1946 and 2009) found a relatively small proportion of left-handers (5.1 percent), half of what one would expect in comparison to the percentage in the general population. This smaller cohort of left-handed players performed well, though, outclassing their right-handed competition in points per game, shooting percentage, assists, and rebounding.[43]

Grouping these sports together by their common use of a fast ball is an oversimplification, though. They also have something else in common, which is that they're all interactive games, where players don't simply interact with the ball but also with other players. Referring to these games as proxies for combat might be going too far, but there are definitely some commonalities between combat sports and these interactive ball games. Another common element in these fast-ball sports is time pressure. In golf, one can take time setting up a shot. In sports such as baseball, cricket, soccer, or tennis, the opponent sets the time frame. There is no waiting for the perfect shot to present itself.

A German study by Florian Loffing examined the relationship between time pressure in sports and the proportion of left-handers among the top 100 ranked players.[44] Time pressure was defined by the mean time interval between the actions of two competing players. For racquet sports, this interval was the time between racquet and ball/birdie contact. Table tennis had very short intervals and therefore high time pressure, whereas badminton or squash were characterized by longer intervals. In sports such as cricket and baseball, the time between the pitch and the contact with the bat was measured to estimate time pressure. Baseball had more time pressure than cricket, and both games were considerably faster than racquet sports, such as tennis or badminton. The fastest sports tended to have the highest proportions of left-handers among their top 100 ranked players, such as baseball (30.39 percent of pitchers), cricket (21.78 percent of bowlers), and table tennis (25.82 percent). Slower interactive ball sports such as squash had the lowest proportion of left-handed players (8.70 percent).

The left-hander advantage in fast-ball sports might also have something to do with a right-handed competitor's inability to "read" or predict the movements of a left-hander accurately. A volleyball study asked this very

Fig. 60: Expert and novice observers watched video clips of the movements preceding a volleyball spike by left- or right-handed attackers. Observers had to predict the direction of the spike based on the preceding movements, and predictions were more accurate for right-handed attackers.

question by presenting video clips leading up to a spike (but stopping the motion before the actual spike) and asking observers to predict the trajectory of the spike (see Fig. 60). Clips of three right-handed and three left-handed volleyball players were presented to 18 skilled and 18 novice observers. Predictions of the outcome of left-handed attacks were less accurate than predictions for right-handed spikes, and the errors were even more pronounced for novice versus experienced observers.[45]

So far our discussion of left/right differences in sport has focused on the lateral preferences of the athletes themselves, and especially on their preferences for one limb or the other. However, there is also an interesting perceptual side to this story. Athletes constantly try to predict where a ball might land, where a defender might be headed, or what offensive tactic is coming from an "attacker" with a ball or fencing foil. Are these judgments and predictions of size, distance, speed, and trajectory also biased by our lopsided brain? Of course, they are!

Let's start with goal-kicking in Australian Rules football. We learned in Chapter 7 that people tend to overestimate the size, number, and proximity of items on the left while underestimating those same things for objects on the right side. Australian Rules football is a particularly convenient game for studying lateral biases in goal-kicking because unlike sports such as hockey or soccer, there is no goalie, and no strategic advantage to try to score on the goalie's weaker side. Goals are often attempted "on the run" but can also be achieved during free kicks (see Fig. 61 for a sample Australian Rules football field). An Australian research group studied goal-kicking attempts from the Australian Football League (AFL), sampled from 16 teams during the 2005–2009 seasons. Knowing that people tend to underestimate the distance and proximity of objects to the right, the researchers predicted more rightward kicks (even the ones that passed between the goalposts) and more goal misses (called a "behind") to the right. They found both of those effects. Players tended to score on the rightward half of the goal and were more likely to miss to the right.[46] The same research group had 212 people attempt to kick a soccer ball in the middle of a goal in the lab. Just as they found in the AFL, amateur soccer kicks tended to deviate toward the right half of the goal.

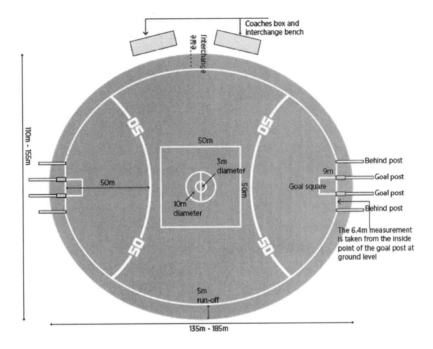

Fig. 61: A typical Australian Rules football field. If the ball passes between the goalposts, a goal is scored. However, if the ball passes between a goalpost and a "behind" post, a "behind" is scored.

It isn't just fast-ball sports that show this rightward bias effect. Golf demonstrates it, too. A study of 30 novice golfers performing 90 putting trials each demonstrated the same type of systemic rightward errors on display in soccer and Australian Rules football.[47] Even experienced archers appear to miss more to the right than they do to the left.[48] Finally, I'm loath to include video games in this chapter on sports, since I am one of the few who still refuses to consider video gaming to be "sport." (Although I happily recognize the many potential benefits of video gaming, including increasing hand-eye coordination, sustained attention, multi-tasking ability, short-term memory, mental health and well-being,[49] and other benefits that I'll keep myself from listing out of fear of compromising my "non-sport stance" on video gaming.) Regardless, the rightward error bias that is evident in "real" sports is also readily observable in video gaming. For example, in the multi-player, first-person shooter game *Counter-Strike: Global Offensive*, players tend to show rightward biases in both fatal assaults and navigational errors.[50]

The vast majority of this chapter has focused on the laterality of the athletes participating in sports. The athletes themselves have gotten all the attention. However, I will end the chapter with a discussion of the sports *observer*. A typical amateur sporting event might have more competitors than spectators, but professional contests can draw millions of spectators and generate even more millions of dollars. Let's consider two types of observers: the ones paid to be there and the ones who pay to be there.

In sports such as tennis, volleyball, basketball, or boxing, the paid observers are the referees, umpires, and linespeople. In sports such as gymnastics, diving, or figure skating, the paid observers actually give the athletes scores for their athletic actions and determine the winners and losers in a somewhat subjective but usually rubric-driven fashion. When runners complete a 100-metre race, the one who wins does so based on really obvious and objective criteria. There are no points for style. But for sports that *do* reward style, do the lefts and rights matter?

One clever study asked this very question within the sport of gymnastics. Testing a Western sample of 48 layperson observers and 48 trained gymnastics evaluators, the researchers presented images or videos of mirror-reversed left-to-right or right-to-left gymnastic movements and asked participants to indicate which movement was more "beautiful" (see Fig. 62). The images and videos were identical except for the direction of the movements. Untrained observers consistently indicated that left-to-right movements were more "beautiful," but trained observers did not exhibit the same left-right bias.

It's reassuring that the trained judges were not biased by the direction of the motion by the gymnasts, but many other sports also involve exhibitions of linear motion and ratings of the beauty of the movement. Even in sports in which the perceived beauty of the movement has no impact on the outcome of the competition, the perceived beauty of the movement is still held in high regard. For people who have ever watched a highlight reel of

Left-Right Directional Movements   Right-Left Directional Movements

Fig. 62: Layperson observers of gymnastic or even non-gymnastic movements indicated that left-right movements were more beautiful than right-left movements, but trained judges didn't demonstrate the same bias.

basketball's best plays of the month, the baskets featured in the reel didn't count any more than any other successful shots that month, but we celebrate them for their beauty, anyway. As layperson observers, are our perceptions of beauty in sport influenced by the direction of the motion?

In soccer, it certainly appears so. In Italy, a research group watched recordings of soccer goals presented in a left-right trajectory or right-left trajectory and rated the goals in terms of beauty ("How beautiful is the goal?"), strength ("How strongly does the player hit the ball?"), and speed ("How fast is the goal?"). All three ratings were higher for the left-right goals, with the movement direction consistent with the native reading direction of the Italian observers.

The same research group also completed a similar study, but instead of presenting potentially "beautiful" sports images, they used aggressive film scenes and asked: "How strongly did the person who hit (or pushed) the other appear?" "How shocked/traumatized did the person who got hit (or pushed) appear?" and "How violent did the scene appear to you?" For all three measures, the Italian observers thought the violent scenes were stronger and more violent when depicting left-right motion compared to right-left motion. When the same study included Arabic-speaking/writing (right-left) participants, the direction of these asymmetries reversed congruent with their reading direction. Arabic-speaking observers thought right-left goals were more "beautiful," and right-left punches and pushes were stronger and more violent.[51]

Collectively, these studies tell us that left/right differences in brain function influence the lateral biases of the players on the field and the perceptions/reactions of the people watching the action. Fortunately, the perceptual biases exhibited by most laypeople seem to be weaker or even absent in trained observers, helping preserve the integrity of judged competitions.

TAKEAWAYS

Sports provide an ideal arena for studying left-right differences. The lateral preferences and performances of athletes are diligently recorded, highly scrutinized and trained, and athletic archives are a treasure trove for laterality researchers. Left-handers clearly excel at many sports, especially those that involve combat (or simulated combat), as well as highly paced fast-ball sports. This advantage can be explained by the fighting hypothesis, that left-handers benefit from an advantage in confrontations of physical aggression, and that this advantage balances any adverse effects that might also accompany left-handedness. The advantage might also be explained by the negative frequency–dependent selection hypothesis in that left-handed people are more successful because they're less numerous (frequent). Finally, the left-right differences are not confined to the athletes themselves. Spectators and even judges are also influenced by left-right perceptual biases in systematic and predictable ways.

Afterword

There are no side effects — only effects. Those
we thought of in advance, the ones we like, we
call the main, or intended, effects, and take credit
for them. The ones we didn't anticipate, the ones
that came around and bit us in the rear — those
are the "side effects."

— JOHN D. STERMAN

We tend to think of side effects as *bad* things. That allergy medicine we take to keep sneezing and sniffles at bay probably also makes us drowsy. Unless we take it right before bed, that side effect isn't beneficial and might actually deter us from using the medication in the first place. In some rare and special circumstances, side effects can also be advantageous. Women take oral contraceptives to prevent unwanted pregnancies, but those drugs also counteract the development of acne. Aspirin is normally used as a painkiller but helps prevent heart attacks and strokes and even leads to higher survival rates in cases of colon or prostate cancer. Minoxidil was developed as a treatment for high blood pressure but is now used topically as a treatment for pattern baldness.

Even side effects that initially *seem* bad can be signs of something good. Writing this during a global pandemic, I cannot resist the temptation to discuss vaccines and their associated side effects. Outside of the most extreme reactions (such as allergic ones like anaphylaxis), the *normal* side effects from vaccines (pain/swelling at the site of injection, fever, nausea, headache, fatigue) are actually signs of the immune response to the vaccine and an indication of its effectiveness. This doesn't mean that side effects are necessary for effective vaccination, but as annoying as they can be, the side effects themselves are normally indicative of something good.

This book obviously has not focused on the benefits and harms of pharmaceuticals, but some of the same principles can apply to the side effects stemming from our lopsided brain. These effects are readily observable but can be so much more than a scientific curiosity or piece of trivia. Instead, we can leverage our emerging knowledge of these side effects to optimize the images we use for our social media, dating profiles, decisions about seat selection, or even advertising campaigns.

I have offered these side effects one at a time, each neatly encapsulated within its own chapter, but there is a downside to organizing the material that way. The reader might get the impression that the lateral biases presented throughout this book are entirely independent of one another, but they are not. Handedness is the most obvious example of a side effect that influences but usually doesn't *cause* other side effects, such as our biases in cradling or preferences for an eye, ear, or foot. In other cases, one side effect can actually cause another. For example, many of the side effects in sports are driven by the handedness of the athletes. Other side effects interact with one another. Our bias toward rightward turning might contribute to the rightward bias in romantic kissing. Our preference for turning the left cheek when posing probably interacts with our preference for leftward lighting. In other words, the dozen side effects I have described in depth and mostly in isolation don't always appear in isolation from one another.

Our most obvious lopsided behaviour is handedness, and we know from analyses of ancient artworks like cave paintings that 90 percent of the human race has been right-handed for more than 50 centuries and that no predominantly left-handed cultures exist. Right-handedness also appears to be uniquely human. Gorillas and monkeys seem to favour the left hand if they exhibit any preference at all, and species such as cats and dogs don't display species-level "paw preferences" the way we tend to prefer the right hand. We also glorify the right (it is synonymous with being correct) while we demonize the left with terms such as *sinister* or *gauche*. Being left-handed runs in families and is associated with many good things (being intellectually gifted, artistic, musical, very good at math). However, it is also linked to a number of bad things (autoimmune disorders, birth stress, schizophrenia, dyslexia). Based on the currently high rates of left-handedness among young people and low rates of left-handedness among the elderly, it can be tempting to conclude that left-handers don't live as long as right-handers. However, factors such as social pressure are responsible for at least some of the disparity.

In addition to preferred hands, we also have strong preferences for feet, ears, and eyes. Unlike our hand preferences, our other lateral preferences are much less obvious to casual observers, which also makes them less vulnerable to social pressures. Many cultures have strict rules about which hand should be used for preparing and consuming food (the right) versus which hand should be employed to clean oneself (the left). However, foot, eye, and ear preferences can develop with much less social interference, and they might provide better clues about an individual's unique brain lateralization. Even though cultural influences on our foot, eye, and ear preferences are less common, most objects are still manufactured for a right-dominated world, including the sights on rifles or microscopes.

Some individuals struggle to differentiate between left and right, and when errors are made, the consequences can range from the mildly annoying to the very serious. For example, swerving in the wrong direction in traffic can be fatal. Left-right surgical mistakes can also be deadly. These errors emerge when using the same simple terms to describe left and right, but matters can become even more confusing when considering the wide variety of terminology utilized to describe our lefts and rights. Many of these terms are value-laden, ascribing positive qualities (straight, correct, accurate, true, clean) to the right, while the left is denigrated with much more negative associations (crooked, wrong, clumsy, false, dirty). Our terminology for differentiating between the right and the left has also entered our political discourse, arising from seating arrangements in the National Constitutive Assembly in France during the time of the French Revolution in the late 18th century.

Kissing is a big deal in popular culture, but the science of smooching has been pretty limited. However, recent research is starting to shed light about how different types of kisses (romantic, parental, social greeting) are expressed differently. When romantic partners kiss, both parties tend to turn to the right. However, when parents kiss their children, this rightward bias disappears. Similarly, if two complete strangers kiss, this lack of familiarity extinguishes the rightward bias. During social-greeting kisses (as is common in Europe), most regions demonstrate a rightward bias, but some parts of the world exhibit leftward biases. Collectively, this research establishes that when performing a romantic kiss, one normally turns to the right, but if kissing a parent or child, central or even leftward kisses are common. When travelling in Europe, look up how many kisses and which cheek to kiss before a first encounter with someone.

Kissing isn't the only "social touch" with side effects, though. More than 2,000 years ago, Plato was among the first to document how people tend to cradle infants to the left. Experienced mothers demonstrate this leftward bias, but so do 15-year-old boys who have never held an infant before, and so do macaque monkeys and gorillas! The first scientific study of this bias came from observations of leftward cradling by monkeys in New York City's Central Park Zoo. Clearly, this is not a bias that we learn through culture or experience, so why do we cradle to the left? One of our more pronounced anatomical asymmetries might be to blame. Except for a few incredibly rare cases of situs inversus, the heart is normally positioned leftward in the chest, and babies are soothed by the sound of a mother's heartbeat. It's also possible that leftward holding creates a greater sense of intimacy and affinity between the parent and child.

Even pictures of single adults demonstrate side effects. When posing for a photo, most people do so to the side, usually turning to the right and presenting the left cheek. This left-cheek bias is obvious whether you survey paintings in a museum (consider famous portraits such as the *Mona Lisa*), flip through a high school yearbook, or even browse (non-mirrored) selfies on Instagram or online dating profiles. Why? Is the left cheek more attractive? We can find clues among images that don't show the leftward bias, such as the profile pictures of famous scientists, teachers (but not students) in high school yearbooks, or even religious leaders. Images of Jesus tend to show the left cheek, but Buddha doesn't appear to turn the left cheek. Showing the left cheek makes individuals appear more emotional and possibly approachable. Centrally posed pictures, like a passport photo, Ministry of Transportation photo, or ID badge at work, tend not to be very flattering. So how should we pose for our next photograph, or which selfie should we choose for the next post? If we want to appear emotional, approachable, and friendly, turn the left cheek. If we desire to appear impassive, objective, and even aloof, choose a more central or even rightward pose. Sometimes posing "right" is posing left.

In addition to this posing bias, famous artwork also tends to be lopsided in another significant way. Over three-quarters of master paintings depict the light source from the left side. We don't need to be a master painter to show the bias, though, since children's drawings show the same leftward bias. It doesn't appear to be caused by the handedness of the artist, or by painting or drawing, because it's even visible in photographs in magazines. Advertisements for products lit from the left get higher product ratings, and consumers are more interested in purchasing them. Side effects are also readily observable in other facets of artistic expression. There are clear lateral biases in the art we produce and reliable lateral biases in how we perceive and react to it. Considerations such as posing direction, lighting direction, centre of mass, direction of movement, and native reading direction of the primary audience should all be taken into account when crafting a painting, plating a meal, or designing a skyscraper.

Other side effects are readily observable in our everyday movements. Hand gestures might be behavioural fossils, left over from a time before we used spoken language to communicate with one another. Some cultures (Italians, for example) appear to express this behavioural fossil more than others (such as the Japanese). The production of language is normally dominated by the left hemisphere of the brain, and when people speak, they tend to make gestures with the right hand (also controlled by the left hemisphere of the brain, even in left-handers). However, when we're listening during a conversation, this lateral bias tends to be reversed. Not only do we make fewer hand gestures, but they are usually left-hand gestures controlled by the right hemisphere.

Another everyday movement demonstrating side effects is head-turning. The tendency to turn one's head to the right is one of the earliest lopsided behaviours in humans. It is clearly visible after 38 weeks of gestation in the fetus, which is long before culture or social learning can influence the child. This bias persists throughout our lives. If we ask an adult to walk down an empty hallway, turn around, and return, chances are he or she will rotate to the right. We can see evidence of this rightward bias when we drive, enter a store, play sports, or even dance. Most ancient dances have circular motions that tend to move in a clockwise (rightward) direction.

We can also observe turning biases when people enter a room and select a seat, but those decisions are more complicated than deciding which direction to turn. What influences the choice of a seat when walking into a classroom or movie theatre? How about selections on seating charts for a big concert or transcontinental flight? The side we choose depends on the type of experience we are expecting and on the lateral biases in our brains. Because the right hemisphere is dominant for processing emotion, and information presented on the left is primarily processed by the right hemisphere, we would expect people to prefer perceiving emotional material from the left. Looking at movie theatre seating preferences, that is exactly what we find. Conversely, the left hemisphere is typically dominant for language processing, and we tend to prefer perceiving language from the right side of space. Looking at classroom seating preferences, that is exactly what we discover. People prefer to sit on the left side of movie theatres and the right side of classrooms. The lateral biases we see for seat selection depend on what people expect to view.

Side effects are also readily observable during other forms of entertainment, such as watching sports. Lateral biases clearly play a big role among players of amateur and professional sports (such as handedness in baseball), and some of the best records of lateral preferences come from athletics.

However, our lateral biases even influence the spectators in sport, determining how beautiful a soccer goal appears to be or how powerful a punch looks during a boxing match. Left-to-right readers tend to prefer watching motion that flows from left to right. Our misperceptions of speed and distance also influence sports performance on the field. We tend to think objects on the left side of space are closer and larger than those on the right.

This book was born, or at least conceived, on a flight home from a conference in Montreal in 2004. I was just starting to explore lighting biases at that time and presented some of those results alongside colleagues working in areas such as cradling biases and posing biases. I left Montreal with my head spinning about how these biases might interact and inform each other and sketched out a rough plan for this book at 9,000 metres. Perhaps it was the thin air or even the normal sleep deprivation that emerges at the end of a good conference, but most of the chapter outlines I sketched out during that flight were only partially viable at that time. Even if I had thrown myself at the project back then, I don't think it could have succeeded. In the years since, many new articles about side effects have emerged (just look at all the references listed in the reference section of this book that have publication dates after 2004). Even entire research groups focusing on these issues have emerged since then, and new and exciting and impactful research in this area is emerging faster than ever. Eventually, there was sufficient quantity, quality, and coherence in the research literature to allow me to actually write this book. It's an exciting time, but also a terrifying one to publish a book like this, knowing that as soon as the ink dries another great new study that just *needs* to be included will appear. Throughout the writing process, I also had to leave a number of chapters on the cutting-room floor (such as "Side Effects in Transportation: Cars, Planes, Boats, and Trains," or "Right Wings to the Left: Side Biases in Politics"), since research in those areas is just starting to build momentum. With a little luck, a sequel and/or update to this book will appear before another 18 years pass.

Acknowledgements

My name might be the only one on the cover of this book, but a project of this scale isn't the work of one person. I have been extremely fortunate to benefit from the support, advice, patience, and even occasional opposition of my family, friends, colleagues, and students.

The team at Dundurn has been wonderful to work with, consistently raising the level of the content I provided, exercising patience and rigour while still propelling the project ahead. My sincerest thanks go out to Russell Smith, Elena Radic, Laura Boyle, Michael Carroll, Kristina Jagger, Farrah Riaz, Kathryn Lane, Sara D'Agostino, Scott Fraser, and the rest of the Dundurn team. My path to Dundurn was not a short or simple one, and I thank Duncan Mackinnon and Deborah Schneider for serving as early advocates for the project.

Thanks to my wife, Lana, daughter, Mileva, and son, Noam, for all their love and support through the project. My parents, John and Alma, have also supported me and the work itself, including comments on early drafts of the work, sage advice, and perhaps most critically of all, encouragement.

At the risk of making the image credits section redundant, I also want to thank the many talented artists for sharing their work on these pages. In some cases (such as images contributed by Mileva Elias), I know the artists well, and they patiently and expertly produced the custom pieces that I requested. In other cases, I reached out to colleagues or strangers near and far, asking for permission to include their work, or sometimes for a new unique image. Thanks to the graciousness and responsiveness of others, this book benefits from pictures of tattoos from Australia, a computer-generated averaged image from thousands of stills taken in Japan, schematic drawings of sports fields, and even famous pieces of artwork, all viewed in a new (probably leftward) light. Given how much of this book is dedicated to discussion about biases in imagery, the inclusion of compelling and illustrative examples of that imagery was critical.

Many of the students who helped support this research through lab-based projects and theses have their names in the references section already, but that list is not exhaustive. I didn't cite every study of a side effect that we ever completed, and, indeed, not every project resulted in a publication at all. The scientific community learned from the projects that were published, but I also learned a lot from the projects that didn't work out. I want to both express my thanks and, ultimately, share the credit for this work with the students I have had the privilege of working with over the years. I have been so thankful for your intellectual curiosity, hard work, and perseverance. I have been proud to watch you develop into professors, clinical psychologists, lawyers, physicians, advertisers, speech pathologists, health care policy analysts, research facilitators, and so much more. My sincerest thanks go out to Abby Holtslander, Alastair MacFadden, Angela Brown, Austen Smith, Brendon Gibson, Brent Robinson, Cathy Burton, Christianne Rooke, Cindy La, Colin Ouellette, Colleen Cochran, Colleen Hardie, Conley Kriegler, Danny Krupp, Elli McDine, Delaine Engebregtson, Dennis Mah, Emma Gardner, Farzana Tessem, Hannah Tran, Izabela Szelest, Jeff Martin, Jennifer Burkitt, Jennifer Hiatt, Jennifer Hutchinson, Jennifer Sedgewick, Jocelyn Poock, Karen Gilleta, Kari Duerksen, Kate Goodall, Katherine McKibbin, Kelly Suschinsky, Kirk Nylen, Laurie Sykes-Tottenham, Leanne Miller, Lisa Lejbak, Lisa Poon, Loni Rhode, Marianne Hrabok, Meghan Flath, Miles Bowman, Miriam Reese, Morsal Niazi, Murray Guylee, Neil Sulakhe, Nicole Thomas, Paula Morton, Punya Miglani, Rebecca Cairns, Regan Patrick, Sarah Simmons,

Sierra Kyliuk, Tamara (Colton) El Hawat, Trista Friedrich, Tyson Baker, and Victoria Harms.

Finally, I wish to thank the academic mentors and colleagues that taught me about our side effects in the first place, opening doors for me to study the topic myself, and supporting my academic path. This path started with Tom Wishart and Margaret Crossley, followed by M.P. Bryden, I.C. McManus, and Barb Bulman-Fleming. Once I returned back to Saskatoon, this work continued with colleagues at the University of Saskatchewan, including Deb Saucier, Scott Bell, Carl Gutwin, and Marla Mickelborough. During this same period, I was inspired by work from Michael Corballis, M.E.R. Nicholls, Gina Grimshaw, Mark McCourt, Michael Peters, Lauren J. Harris, Sebastian Ocklenburg, Julian Packheiser, and Matia Okubo.

So, despite my name standing alone on this book's cover, I share the credit for this work with a long list of wonderful people, only some of whom are featured in this section.

References

INTRODUCTION

1. Stanley Coren and Clare Porac, "Fifty Centuries of Right-Handedness: The Historical Record," *Science* 198, no. 4317 (1977): 631–32.

2. Lealani Mae Y. Acosta, John B. Williamson, and Kenneth M. Heilman, "Which Cheek Did Jesus Turn?" *Religion, Brain & Behavior* 3, no. 3 (2013): 210–18.

3. Avery N. Gilbert and Charles J. Wysocki, "Hand Preference and Age in the United States," *Neuropsychologia* 30, no. 7 (July 1992): 601–08, https://doi.org/10.1016/0028-3932(92)90065-T.

4. Juhn Wada, Robert Clarke, and Anne Hamm, "Cerebral Hemispheric Asymmetry in Humans: Cortical Speech Zones in 100 Adult and 100 Infant Brains," *Archives of Neurology* 32, no. 4 (April 1975): 239–46, https://doi.org/10.1001/archneur.1975.00490460055007.

5. Robin Weatherill et al., "Is Maternal Depression Related to Side of Infant Holding?" *International Journal of Behavioral Development* 28, no. 5 (2004): 421–27.

6. Lorin J. Elias and Deborah M. Saucier, *Neuropsychology: Clinical and Experimental Foundations* (Boston: Pearson/Allyn & Bacon, 2006).

7. Elias and Saucier, *Neuropsychology*.

8. Tino Stöckel and David P. Carey, "Laterality Effects on Performance in Team Sports: Insights from Soccer and Basketball," in *Laterality in Sports: Theories and Applications*, eds. Florian Loffing et al. (London: Elsevier/Academic Press, 2016), 309–28.

9. Thomas R. Barrick et al., "Automatic Analysis of Cerebral Asymmetry: An Exploratory Study of the Relationship Between Brain Torque and Planum Temporale Asymmetry," *NeuroImage* 24, no. 3 (February 1, 2005): 678–91.

10. Norman Geschwind and Walter Levitsky, "Human Brain: Left-Right Asymmetries in Temporal Speech Region," *Science* 161, no. 3837 (July 12, 1968): 186–87.

11. Elias and Saucier, *Neuropsychology.*

12. Marc H.E. de Lussanet, "Opposite Asymmetries of Face and Trunk and of Kissing and Hugging, as Predicted by the Axial Twist Hypothesis," *PeerJ* 7, no. e7096 (June 7, 2019).

13. Elias and Saucier, *Neuropsychology.*

14. Geoffrey J.M. Parker et al., "Lateralization of Ventral and Dorsal Auditory-Language Pathways in the Human Brain," *NeuroImage* 24, no. 3 (February 1, 2005): 656–66.

CHAPTER 1 HANDEDNESS: ARE LEFT-HANDERS ALWAYS RIGHT?

1. Raymond A. Dart, "The Predatory Implemental Technique of Australopithecus," *American Journal of Physical Anthropology* 7, no. 1 (March 1949): 1–38.

2. Nicholas Toth, "Archaeological Evidence for Preferential Right-Handedness in the Lower and Middle Pleistocene, and Its Possible Implications," *Journal of Human Evolution* 14, no. 6 (September 1985): 607–14.

3. Davidson Black, Pierre Teilhard de Chardin, C.C. Young, and W.C. Pei, *Fossil Man in China: The Choukoutien Cave Deposits with a Synopsis of Our Present Knowledge of the Late Cenozoic in China* (New York: AMS Press, 1933).

4. H.W. Magoun, "Discussion of Brain Mechanisms in Speech," in *Brain Function: Speech, Language, and Communication*, ed. Edward C. Carterette (Los Angeles: University of California Press, 1966).

5. Daniel G. Brinton, "Left-Handedness in North American Aboriginal Art," *American Anthropologist* 9, no. 5 (May 1896): 175–81.

6. Wayne Dennis, "Early Graphic Evidence of Dextrality in Man," *Perceptual and Motor Skills* 8, no. 2 (September 1958): 147–49, https://doi.org/10.2466/pms.1958.8.h.147.

7. Coren and Porac, "Fifty Centuries of Right-Handedness."

8. Coren and Porac, "Fifty Centuries of Right-Handedness."

9. I.C. McManus, "The History and Geography of Human Handedness," in *Language Lateralization and Psychosis*, eds. Iris E.C. Sommer and René S. Kahn (Cambridge: Cambridge University Press, 2009), 37–58.

10. Gilbert and Wysocki, "Hand Preference and Age in the United States."

11. McManus, "The History and Geography of Human Handedness."

12. Chris McManus, "Half a Century of Handedness Research: Myths, Truths; Fictions, Facts; Backwards, but Mostly Forwards," *Brain and Neuroscience Advances* 3, nos. 1–10 (2019), doi.org/10.1177/2398212818820513.

13. Kenneth Hugdahl, Paul Satz, Maura Mitrushina, and Eric N. Miller, "Left-Handedness and Old Age: Do Left-Handers Die Earlier?" *Neuropsychologia* 31, no. 4 (April 1993): 325–33.

14. Joseph L. Reichler, ed., *The Baseball Encyclopedia* (New York: Macmillan, 1979).

15. Stanley Coren and Diane Halpern, "Left-Handedness: A Marker for Decreased Survival Fitness," *Psychological Bulletin* 109, no. 1 (1991): 90–106.

16. Diane Halpern and Stanley Coren, "Left-Handedness and Life Span: A Reply to Harris," *Psychological Bulletin* 114, no. 2 (1993): 235–41.

17. John P. Aggleton, J. Martin Bland, Robert W. Kentridge, and Nicholas J. Neave, "Handedness and Longevity: Archival Study of Cricketers," BMJ 309, no. 6970 (1994): 1681–84.

18. Hicks et al., "Do Right-Handers Live Longer? An Updated Assessment of Baseball Player Data," *Perceptual and Motor Skills* 78, nos. 1243–47, https://doi.org/10.2466/pms.1994.78.3c.1243.

19. Per-Gunnar Persson and Peter Allebeck, "Do Left-Handers Have Increased Mortality?" *Epidemiology* 5, no. 3 (May 1994): 337–40.

20. Tyler P. Lawler and Frank H. Lawler, "Left-Handedness in Professional Basketball: Prevalence, Performance, and Survival," *Perceptual and Motor Skills* 113, no. 3 (December 2012): 815–24.

21. James R. Cerhan, Aaron R. Folsom, John D. Potter, and Ronald J. Prineas, "Handedness and Mortality Risk in Older Women," *American Journal of Epidemiology* 140, no. 4 (1994: 368–74.

22. Hugdahl, Satz, Mitrushina, and Miller, "Left-Handedness and Old Age."

23. Lauren J. Harris, "Left-Handedness and Life Span: Reply to Halpern and Coren," *Psychological Bulletin* 114, no. 2 (1993): 242–47.

24. Yukihide Ida, Tanusree Dutta, and Manas K. Mandal, "Side Bias and Accidents in Japan and India," *International Journal of Neuroscience* 111, nos. 1–2 (January 2001): 89–98.

25. Maharaj Singh and M.P. Bryden, "The Factor Structure of Handedness in India," *International Journal of Neuroscience* 74, nos. 1–4 (January-February 1994): 33–43, https://doi.org/10.3109/00207459408987227.

26. Clare Porac, Laura Rees, and Terri Buller, "Switching Hands: A Place for Left Hand Use in a Right Hand World," in *Left-Handedness: Behavioral Implications and Anomalies*, ed. Stanley Coren (Amsterdam: North-Holland, 1990), 259–90.

27. McManus, "The History and Geography of Human Handedness."

28. Ian Christopher McManus and M.P. Bryden, "The Genetics of Handedness, Cerebral Dominance, and Lateralization," in *Handbook of Neuropsychology*, vol. 6, eds., François Boller and Jordan Grafman (Amsterdam: Elsevier, 1992), 115–44.

29. Louise Carter-Saltzman, "Biological and Sociocultural Effects on Handedness: Comparison Between Biological and Adoptive Families," *Science* 209, no. 4462 (1980): 1263–65.

30. Robert E. Hicks and Marcel Kinsbourne, "On the Genesis of Human Handedness," *Journal of Motor Behavior* 8, no. 4 (1976): 257–66, https://doi.org/10.1080/00222895.1976.10735080.

31. Curtis Hardyck and Lewis F. Petrinovich, "Left-Handedness," *Psychological Bulletin* 84, no. 3 (1977): 385–404.

32. McManus and Bryden, "The Genetics of Handedness, Cerebral Dominance, and Lateralization."

33. John Jackson, *Ambidexterity or Two-Handedness and Two Brainedness* (London: Kegan Paul, Trench, Trübner, 1905).

34. Abram Blau, *The Master Hand: A Study of the Origin and Meaning of Left and Right Sidedness and Its Relation to Personality and Language* (New York: American Orthopsychiatric Association, 1946).

35. Carter-Saltzman, "Biological and Sociocultural Effects on Handedness."

36. Hicks and Kinsbourne, "On the Genesis of Human Handedness."

37. Coren and Porac, "Fifty Centuries of Right-Handedness."

38. Lena Sophie Pfeifer et al., "Handedness in Twins: Meta-Analyses" (March 2021): 1–49, https://doi.org/10.31234/osf.io/gy2nx.

39. Michael Reiss et al., "Laterality of Hand, Foot, Eye, and Ear in Twins," *Laterality* 4, no. 3 (July 1999): 287–97.

40. Peter J. Hepper, "The Developmental Origins of Laterality: Fetal Handedness," *Developmental Psychobiology* 55, no. 6 (September 2013): 588–95.

41. Angelo Bisazza, L.J. Rogers, and Giorgio Vallortigara, "The Origins of Cerebral Asymmetry: A Review of Evidence of Behavioural and Brain Lateralization in Fishes, Reptiles and Amphibians," *Neuroscience and Biobehavioral Reviews* 22, no. 3 (1998): 411–26.

42. Lauren Julius Harris, "Left-Handedness: Early Theories, Facts, and Fancies," in *Neuropsychology of Left-Handedness*, ed. Jeannine Herron (Toronto: Academic Press, 1980), 3–78.

43. Lauren Julius Harris, "In Fencing, Are Left-Handers Trouble for Right-Handers? What Fencing Masters Said in the Past and What Scientists Say Today," in *Laterality in Sports: Theories and Applications*, eds. Florian Loffing et al. (London: Elsevier/Academic Press, 2016), 50.

44. Coren and Porac, "Fifty Centuries of Right-Handedness."

45. David W. Frayer et al., "OH-65: The Earliest Evidence for Right-Handedness in the Fossil Record," *Journal of Human Evolution* 100 (November 2016): 65–72.

46. McManus, "The History and Geography of Human Handedness."

47. Johan Torgersen, "Situs Inversus, Asymmetry, and Twinning," *American Journal of Human Genetics* 2, no. 4 (December 1950): 361–70.

48. E.A. Cockayne. "The Genetics of Transposition of the Viscera," *QJM: An International Journal of Medicine* 7, no. 3 (1938): 479–93, https://doi.org/10.1093/oxfordjournals.qjmed.a068598.

49. Torgersen, "Situs Inversus, Asymmetry, and Twinning."

50. Lauren Julius Harris, "Side Biases for Holding and Carrying Infants: Reports from the Past and Possible Lessons for Today," *Laterality* 15, nos. 1–2 (2010): 56–135.

51. Sebastian Ocklenburg et al., "Hugs and Kisses: The Role of Motor Preferences and Emotional Lateralization for Hemispheric Asymmetries in Human Social Touch," *Neuroscience & Biobehavioral Reviews* 95 (December 2018) 95: 353–60.

52. Stanley Coren, *The Left-Hander Syndrome: The Causes & Consequences of Left-Handedness* (New York: The Free Press, 1992).

53. Norman Geschwind and Albert M. Galaburda, "Cerebral Lateralization: Biological Mechanisms, Associations, and Pathology: III. A Hypothesis and

a Program for Research," *Archives of Neurology* 42, no. 7 (1985): 634–54.

54. Sunil Vasu Kalmady et al., "Revisiting Geschwind's Hypothesis on Brain Lateralisation: A Functional MRI Study of Digit Ratio (2D:4D) and Sex Interaction Effects on Spatial Working Memory," *Laterality* 18, no. 5 (2013): 625–40.

55. Elias and Saucier, *Neuropsychology.*

56. Gina M. Grimshaw, Philip M. Bryden, and Jo-Anne K. Finegan, "Relations Between Prenatal Testosterone and Cerebral Lateralization in Children," *Neuropsychology* 9, no. 1 (1995): 68–79.

57. Paul Bakan, Gary Dibb, and Phil Reed, "Handedness and Birth Stress," *Neuropsychologia* 11, no. 3 (July 1973): 363–66.

58. Paul Satz, Donna L. Orsini, Eric Saslow, and Rolando Henry, "The Pathological Left-Handedness Syndrome," *Brain and Cognition* 4, no. 1 (January 1985): 27–46.

59. Elias and Saucier, *Neuropsychology.*

60. Murray Schwartz, "Handedness, Prenatal Stress and Pregnancy Complications," *Neuropsychologia* 26, no. 6 (1988): 925–29.

61. Gail Ross, Evelyn Lipper, and Peter A.M. Auld, "Hand Preference, Prematurity and Developmental Outcome at School Age," *Neuropsychologia* 30, no. 5 (May 1992): 483–94.

62. Alise A. van Heerwaarde et al., "Non-Right-Handedness in Children Born Extremely Preterm: Relation to Early Neuroimaging and Long-Term Neurodevelopment," *PLoS ONE* 15, no. 7 (July 6, 2020): 1–17, http://dx.doi.org/10.1371/journal.pone.0235311.

63. Jacqueline Fagard et al., "Is Handedness at Five Associated with Prenatal Factors?" *International Journal of Environmental Research and Public Health* 18, no. 7 (April 2021): 1–24.

64. Elias and Saucier, *Neuropsychology.*

65. Christopher S. Ruebeck, Joseph E. Harrington, and Robert Moffitt, "Handedness and Earnings," *Laterality* 12, no. 2 (2007): 101–20.

66. H.H. Newman, "Studies of Human Twins: II. Asymmetry Reversal, of Mirror Imaging in Identical Twins," *The Biological Bulletin* 55, no. 4 (1928): 298–315.

67. Salvator Levi, "Ultrasonic Assessment of the High Rate of Human Multiple Pregnancy in the First Trimester," *Journal of Clinical Ultrasound* 4, no. 1 (February 1976): 3–5.

68. Helain J. Landy and L.G. Keith, "The Vanishing Twin: A Review," *Human Reproduction Update* 4, no. 2 (1998): 177–83.

69. Landy and Keith, "The Vanishing Twin."

70. Gregory V. Jones and Maryanne Martin, "Seasonal Anisotropy in Handedness," *Cortex* 44, no. 1 (January 2008): 8–12.

71. Ramon M. Cosenza and Sueli A. Mingoti, "Season of Birth and Handedness Revisited," *Perceptual and Motor Skills* 81, no. 2 (October 1995): 475–80.

72. Georges Dellatolas, Florence Curt, and Joseph Lellouch, "Birth Order and Month of Birth Are Not Related with Handedness in a Sample of 9,370 Young Men," *Cortex* 27, no. 1 (March 1991): 137–40, http://dx.doi.org/10.1016/S0010-9452(13)80277-8.

73. Nathlie A. Badian, "Birth Order, Maternal Age, Season of Birth, and Handedness," *Cortex* 19, no. 4 (December 1983): 451–63, http://dx.doi.org/10.1016/S0010-9452(83)80027-6.

74. Ulrich S. Tran, Stefan Stieger, and Martin Voracek, "Latent Variable Analysis Indicates That Seasonal Anisotropy Accounts for the Higher Prevalence of Left-Handedness in Men," *Cortex* 57 (August 2014): 188–97.

75. Carolien de Kovel, Amaia Carrión-Castillo, and Clyde Francks, "A Large-Scale Population Study of Early Life Factors Influencing Left-Handedness," *Scientific Reports* 9, no. 584 (January 2019): 1–11.

76. Fagard et al., "Is Handedness at Five Associated with Prenatal Actors?"

77. Coren and Halpern, "Left-Handedness."

CHAPTER 2 FEET, EYES, EARS, NOSES: STARTING ON THE RIGHT FOOT

1. Lorin J. Elias and M.P. Bryden, "Footedness Is a Better Predictor of Language Lateralisation Than Handedness," *Laterality* 3, no. 1 (1998): 41–52.

2. Stöckel and Carey, "Laterality Effects on Performance in Team Sports."

3. Nikitas Polemikos and Christine Papaeliou, "Sidedness Preference as an Index of Organization of Laterality," *Perceptual and Motor Skills* 91, no. 3, part 2 (December 2000): 1083–90.

4. Stanley Coren, "The Lateral Preference Inventory for Measurement of Handedness, Footedness, Eyedness, and Earedness: Norms for Young Adults," *Bulletin of the Psychonomic Society* 31, no. 1 (1993): 1–3.

5. Elias and Bryden, "Footedness Is a Better Predictor of Language Lateralisation Than Handedness."

6. Till Utesch, Stjin Valentijn Mentzel, Bernd Strauss, and Dirk Büsch, "Measurement of Laterality and Its Relevance for Sports," in *Laterality in Sports: Theories and Applications*, eds. Florian Loffing et al. (London: Elsevier/Academic Press, 2016), 65–86.

7. Sacco et al., "Joint Assessment of Handedness and Footedness Through Latent Class Factor Analysis," *Laterality* 23, no. 6 (November 2018): 643–63.

8. Elias and Bryden, "Footedness Is a Better Predictor of Language Lateralisation Than Handedness."

9. Lainy B. Day and Peter F. MacNeilage, "Postural Asymmetries and Language Lateralization in Humans (*Homo sapiens*)," *Journal of Comparative Psychology* 110, no. 1 (1996): 88–96.

10. A. Mark Smith, "Giambattista Della Porta's Theory of Vision in the *De refractione* of 1593: Sources, Problems, Implications," in *The Optics of Giambattista Della Porta (ca. 1535–1615): A Reassessment*, eds. Arianna Borelli, Giora Hon, and Yaakov Zik (New York: Springer, 2017), 97–123, http://link.springer.com/10.1007/978-3-319-50215-1_5.

11. D.C. Bourassa, Ian Christopher McManus, and M.P. Bryden, "Handedness and Eye-Dominance: A Meta-Analysis of Their Relationship," *Laterality* 1, no. 1 (March 1996): 5–34.

12. Michael Reiss, "Ocular Dominance: Some Family Data," *Laterality* 2, no. 1 (1997): 7–16.

13. Polemikos and Papaeliou, "Sidedness Preference as an Index of Organization of Laterality."

14. Coren, "The Lateral Preference Inventory for Measurement of Handedness, Footedness, Eyedness, and Earedness."

15. Elias and Saucier, *Neuropsychology*.

16. Giovanni Berlucchi and Salvatore Aglioti, "Interhemispheric Disconnection Syndromes," in *Handbook of Clinical and Experimental Neuropsychology*, eds. Gianfranco Denes and Luigi Pizzamiglio (Hove, United Kingdom: Psychology Press, 1999), 635–70.

17. S.L. Youngentob et al., "Olfactory Sensitivity: Is There Laterality?" *Chemical Senses* 7, no. 1 (January 1982): 11–21, https://doi.org/10.1093/chemse/7.1.11.

18. Youngentob et al. ,"Olfactory Sensitivity."

19. Richard E. Frye, Richard L. Doty, and Paul Shaman, "Bilateral and Unilateral Olfactory Sensitivity: Relationship to Handedness and Gender," in *Chemical Signals in Vertebrates 6*, eds. Richard L. Doty and Dietland Müller-Schwarze (New York: Springer, 1992), 559–64.

20. Moustafa Bensafi et al., "Perceptual, Affective, and Cognitive Judgments of Odors: Pleasantness and Handedness Effects," *Brain and Cognition* 51, no. 3 (2003): 270–75.

21. Robert J. Zatorre and Marilyn Jones-Gotman, "Right-Nostril Advantage for Discrimination of Odors," *Perception & Psychophysics* 47, no. 6 (1990): 526–31.

22. Thomas Hummel, Par Mohammadian, and G. Kobal, "Handedness Is a Determining Factor in Lateralized Olfactory Discrimination," *Chemical Senses* 23, no. 5 (October 1998): 541–44.

23. Richard Kayser, "Luftdurchgangigkeit der Nase," *Archives of Laryngology and Rhinology* 3 (1895): 101–20.

24. Alfonso Luca Pendolino, Valerie J. Lund, Ennio Nardello, and Giancarlo Ottaviano, "The Nasal Cycle: A Comprehensive Review," *Rhinology Online* 1, no. 1 (June 2018): 67–76, http://doi.org/10.4193/RHINOL/18.021.

25. Alan Searleman, David E. Hormung, Emily Stein, and Leah Brzuskiewicz, "Nostril Dominance: Differences in Nasal Airflow and Preferred Handedness," *Laterality* 10, no. 2 (April 2005): 111–20.

26. Raymond M. Klein, David Pilon, Susan Marie Prosser, and David Shannahoff-Khalsa, "Nasal Airflow Asymmetries and Human Performance," *Biological Psychology* 23, no. 2 (1986): 127–37.

27. Susan A. Jella and David Shannahoff-Khalsa, "The Effects of Unilateral Forced Nostril Breathing on Cognitive Performance," *International Journal of Neuroscience* 73, nos. 1–2 (1993): 61–68.

28. Deborah M. Saucier, Farzana Karim Tessem, Aaron H. Sheerin, and Lorin Elias, "Unilateral Forced Nostril Breathing Affects Dichotic Listening for Emotional Tones," *Brain and Cognition* 55, no. 2 (July 2004): 403–05.

29. Robert Hertz, "The Pre-Eminence of the Right Hand: A Study in Religious Polarity," reprint translated by Rodney and Claudia Needham, *HAU: Journal of Ethnographic Theory* 3, no. 2 (2013): 335–57.

CHAPTER 3 WORDS: THE LEFT ISN'T TREATED RIGHT

1. Alan Cienki, "The Strengths and Weaknesses of the Left/Right Polarity in Russian: Diachronic and Synchronic Semantic Analyses," in *Issues in Cognitive Linguistics: 1993 Proceedings of the International Cognitive Linguistics Conference*, eds. Leon de Stadler and Christoph Eyrich (Berlin: De Gruyter Mouton, 1999), 299–330, https://doi.org/10.1515/9783110811933.299.

2. Alan Cienki, "STRAIGHT: An Image Schema and Its Metaphorical Extensions," *Cognitive Linguistics* 9, no. 2 (January 1998): 107–49.

3. H. Julia Hannay, P.J. Ciaccia, Joan W. Kerr, and Darlene Barrett, "Self-Report of Right-Left Confusion in College Men and Women," *Perceptual and Motor Skills* 70, no. 2 (April 1990): 451–57.

4. Sebastian Ocklenburg, "Why Do I Confuse Left and Right?" *Psychology Today*, March 9, 2019, psychologytoday.com/ca/blog/the-asymmetric-brain/201903/why-do-i-confuse-left-and-right.

5. Sonja H. Ofte and Kenneth Hugdahl, "Right-Left Discrimination in Male and Female, Young and Old Subjects," *Journal of Clinical and Experimental Neuropsychology* 24, no. 1 (February 2002): 82–92.

6. Ineke J.M. van der Ham, H. Chris Dijkerman, and Haike E. van Stralen, "Distinguishing Left from Right: A Large-Scale Investigation of Left-Right Confusion in Healthy Individuals," *Quarterly Journal of Experimental Psychology* 74, no. 3 (2021): 497–509, https://doi.org/10.1177/1747021820968519.

7. Ad Foolen, "The Value of Left and Right," in *Emotion in Discourse*, eds., J. Lachlan Mackenzie and Laura Alba-Juez (Amsterdam: John Benjamins Publishing, 2019), 139–58.

8. Cienki, "The Strengths and Weaknesses of the Left/Right Polarity in Russian," 299–330.

9. Cienki, "The Strengths and Weaknesses of the Left/Right Polarity in Russian," 299–330.

10. Cienki, "The Strengths and Weaknesses of the Left/Right Polarity in Russian," 299–330.

11. Juanma de la Fuente, Daniel Casasanto, Antonio Román, and Julio Santiago, "Searching for Cultural Influences on the Body-Specific Association of Preferred Hand and Emotional Valence," *Proceedings of the*

33rd Annual Conference of the Cognitive Science Society 33 (July 2011): 2616–20, https://cloudfront.escholarship.org/dist/prd/content/qt6qc0z1zp/qt6qc0z1zp.pdf.

12. Juanma de la Fuente, Daniel Casasanto, Antonio Román, and Julio Santiago, "Can Culture Influence Body-Specific Associations Between Space and Valence?" *Cognitive Science* 39, no. 4 (May 2015: 821–32, http://doi.wiley.com/10.1111/cogs.12177.

13. Foolen, "The Value of Left and Right."

14. Wulf Schiefenhövel, "Biased Semantics for Right and Left in 50 Indo-European and Non-Indo-European Languages," *Annals of the New York Academy of Sciences* 1288, no. 1 (June 2013): 135–52.

15. Foolen, "The Value of Left and Right."

16. Foolen, "The Value of Left and Right."

17. Schiefenhövel, "Biased Semantics for Right and Left in 50 Indo-European and Non-Indo-European Languages."

18. Lorin J. Elias, "Secular Sinistrality: A Review of Popular Handedness Books and World Wide Web Sites," *Laterality* 3, no. 3 (1998): 193–208.

19. Simon Langford, *The Left-Handed Book: How to Get By in a Right-Handed World* (London: Panther, 1984).

20. Leigh W. Rutledge and Richard Donley, *The Left-Hander's Guide to Life: A Witty and Informative Tour* (New York: Plume/Penguin, 1992).

21. Daniel Casasanto, "Embodiment of Abstract Concepts: Good and Bad in Right- and Left-Handers," *Journal of Experimental Psychology: General* 138, no. 3 (August 2009): 351–67.

22. Daniel Casasanto and Kyle Jasmin, "Good and Bad in the Hands of Politicians: Spontaneous Gestures During Positive and Negative Speech," *PLoS ONE* 5, no. 7 (July 28, 2010).

23. John T. Jost, "Elective Affinities: On the Psychological Bases of Left-Right Differences," *Psychological Inquiry* 20, nos. 2–3 (April 2009): 129–41.

CHAPTER 4 KISSING: ARE WE DOING IT RIGHT?

1. J. Ridley Stroop, "Studies in Interference in Serial Verbal Reactions," *Journal of Experimental Psychology* 18, no. 6 (1935): 643–62.

2. Antina de Boer, E.M. van Buel, and Gert J. ter Horst, "Love Is More Than Just a Kiss: A Neurobiological Perspective on Love and Affection,"

Neuroscience 201 (January 10, 2012): 114–24, http://dx.doi.org/10.1016/j. neuroscience.2011.11.017.

3.   Helen Fisher, Arthur Aron, and Lucy L. Brown, "Romantic Love: An fMRI Study of a Neural Mechanism for Mate Choice," *The Journal of Comparative Neurology* 493, no. 1 (December 2005): 58–62.

4.   Sheril Kirshenbaum, *The Science of Kissing: What Our Lips Are Telling Us* (New York: Grand Central Publishing, 2011).

5.   Onur Güntürkün, "Adult Persistence of Head-Turning Asymmetry" *Nature*, 421, (2003): 711.

6.   Güntürkün, "Adult Persistence of Head-Turning Asymmetry."

7.   Dianne Barrett, Julian G. Greenwood, and John F. McCullagh, "Kissing Laterality and Handedness," *Laterality* 11, no. 6 (November 2006): 573–79.

8.   John van der Kamp and Rouwen Cañal-Bruland, "Kissing Right? On the Consistency of the Head-Turning Bias in Kissing," *Laterality* 16, no. 3 (May 2011): 257–67.

9.   Julian Packheiser et al., "Embracing Your Emotions: Affective State Impacts Lateralisation of Human Embraces," *Psychological Research* 83, no. 1 (February 2019): 26–36.

10.   Samuel Shaki, "What's in a Kiss? Spatial Experience Shapes Directional Bias During Kissing," *Journal of Nonverbal Behavior* 37, no. 1 (2013): 43–50.

11.   Sedgewick, Holtslander, and Lorin J. Elias, "Kissing Right? Absence of Rightward Directional Turning Bias During First Kiss Encounters Among Strangers," *Journal of Nonverbal Behavior* (2019).

12.   Jennifer Sedgewick and Lorin J. Elias, "Family Matters: Directionality of Turning Bias While Kissing Is Modulated by Context," *Laterality* 21, nos. 4–6 (July-November 2016): 662–71, http://dx.doi.org/10.1080/1357650X .2015.1136320.

13.   Barrett, Greenwood, and McCullagh, "Kissing Laterality and Handedness."

14.   Jacqueline Liederman and Marcel Kinsbourne, "Rightward Motor Bias in Newborns Depends Upon Parental Right-Handedness," *Neuropsychologia* 18, nos. 4–5 (1980): 579–84.

15.   Andreas Bartels and Semir Zeki, "Neural Basis of Love," *NeuroReport* 11, no. 17 (2000): 3829–34.

16.   Andreas Bartels and Semir Zeki, "The Neural Correlates of Maternal and Romantic Love," *NeuroImage* 21, no. 3 (March 2004): 1155–66.

17. Sedgewick and Elias, "Family Matters."

18. Sedgewick, Holtslander, and Elias, "Kissing Right?"

19. Ryan S. Elder and Aradhna Krishna, "The 'Visual Depiction Effect' in Advertising: Facilitating Embodied Mental Simulation Through Product Orientation," *Journal of Consumer Research* 38, no. 6 (April 2012): 988–1003.

20. Sedgewick, Holtslander, and Elias, "Kissing Right?"

21. Shaki, "What's in a Kiss?"

22. Amandine Chapelain et al., "Can Population-Level Laterality Stem from Social Pressures? Evidence from Cheek Kissing in Humans," *PLoS ONE* 10, no. 8 (2015): e0124477, http://dx.doi.org/10.1371/journal.pone.0124477.

23. Chapelain et al., "Can Population-Level Laterality Stem from Social Pressures? Evidence from Cheek Kissing in Humans."

24. Chapelain et al., "Can Population-Level Laterality Stem from Social Pressures? Evidence from Cheek Kissing in Humans."

CHAPTER 5 CRADLING BIASES: ARE YOU HOLDING YOUR BABY RIGHT?

1. Blau, *The Master Hand*.

2. Plato, *The Laws of Plato*, trans. Thomas L. Pangle (Chicago: University of Chicago Press, 1988).

3. Harris, "Side Biases for Holding and Carrying Infants," 64.

4. Harris, "Side Biases for Holding and Carrying Infants," 64.

5. Harris, "Side Biases for Holding and Carrying Infants," 73.

6. Harris, "Side Biases for Holding and Carrying Infants," 74.

7. Jean-Jacques Rousseau, *Confessions*, ed. Patrick Coleman, trans. Angela Scholar (Oxford: Oxford University Press, 2008).

8. Lee Salk, "The Role of the Heartbeat in the Relations Between Mother and Infant," *Scientific American* 228, no. 5 (May 1973): 24–29.

9. Salk, "The Role of the Heartbeat in the Relations Between Mother and Infant," 24.

10. Harris, "Side Biases for Holding and Carrying Infants," 57.

11. Stanley Finger, "Child-Holding Patterns in Western Art," *Child Development* 46, no. 1 (1975): 267–71.

12. G. Alvarez, "Child-Holding Patterns and Hemispheric Bias," *Ethology and Sociobiology* 11, no. 2 (1990): 75–82.

13. Lauren Julius Harris, Rodrigo A. Cárdenas, Nathaniel D. Stewart, and Jason B. Almerigi, "Are Only Infants Held More Often on the Left? If So, Why? Testing the Attention-Emotion Hypothesis with an Infant, a Vase, and Two Chimeric Tests, One 'Emotional,' One Not," *Laterality* 24, no. 1 (January 2019): 65–97.

14. Masayuki Nakamichi, "The Left-Side Holding Preference Is Not Universal: Evidence from Field Observations in Madagascar," *Ethology and Sociobiology* 17, no. 3 (May 1996): 173–79.

15. C.U.M. Smith, "Cardiocentric Neurophysiology: The Persistence of a Delusion," *Journal of the History of the Neurosciences* 22, no. 1 (2013): 6–13.

16. John Patten, *Neurological Differential Diagnosis*, 2nd ed. (New York: Springer, 1996).

17. D.N. Kennedy et al., "Structural and Functional Brain Asymmetries in Human Situs Inversus Totalis," *Neurology* 53, no. 6 (October 1999): 1260–65.

18. Salk, "The Role of the Heartbeat in the Relations Between Mother and Infant," 29.

19. Brenda Todd and George Butterworth, "Her Heart Is in the Right Place: An Investigation of the 'Heartbeat Hypothesis' as an Explanation of the Left Side Cradling Preference in a Mother with Dextrocardia," *Early Development and Parenting* 7, no. 4 (2002): 229–33.

20. Salk, "The Role of the Heartbeat in the Relations Between Mother and Infant."

21. I. Hyman Weiland, "Heartbeat Rhythm and Maternal Behavior," *Journal of the American Academy of Child Psychiatry* 3, no. 1 (January 1964): 161–64.

22. Harris, Cárdenas, Stewart, and Almerigi, "Are Only Infants Held More Often on the Left?"

23. Harris, Cárdenas, Stewart, and Almerigi, "Are Only Infants Held More Often on the Left?"

24. I. Hyman Weiland and Zanwil Sperber, "Patterns of Mother-Infant Contact: The Significance of Lateral Preference," *The Journal of Genetic Psychology* 117, no. 2 (December 1970): 157–65, https://doi.org/10.1080/00221325.1970.10532575.

25. Ernest L. Abel, "Human Left-Sided Cradling Preferences for Dogs," *Psychological Reports* 107, no. 1 (August 2010): 336–38.

26. Dale Dagenbach, Lauren Julius Harris, and Hiram E. Fitzgerald, "A Longitudinal Study of Lateral Biases in Parents' Cradling and Holding of Infants," *Infant Mental Health Journal* 9, no. 3 (Fall 1988): 218–34, https://vdocuments.net/reader/full/a-longitudinal-study-of-lateral-biases-in-parents-cradling-and-holding-of.

27. Joan S. Lockard, Paul C. Daley, and Virginia M. Gunderson, "Maternal and Paternal Differences in Infant Carry: U.S. and African Data," *The American Naturalist* 113, no. 2 (February 1979): 235–46.

28. Peter de Château, "Left-Side Preference in Holding and Carrying Newborn Infants: A Three-Year Follow-Up Study," *Acta Psychiatrica Scandinavica* 75, no. 3 (March 1987): 283–86, https://doi.org/10.1111/j.1600-0447.1987.tb02790.x.

29. Peter de Château, M. Mäki, and B. Nyberg, "Left-Side Preference in Holding and Carrying Newborn Infants III: Mothers' Perception of Pregnancy One Month Prior to Delivery and Subsequent Holding Behaviour During the First Postnatal Week," *Journal of Psychosomatic Obstetrics & Gynecology* 1, no. 2 (1982): 72–76.

30. de Château, Mäki, and Nyberg, "Left-Side Preference in Holding and Carrying Newborn Infants III."

31. Weatherill et al., "Is Maternal Depression Related to Side of Infant Holding?"

32. Paul Richter, Andrés Hseerlein, Hermes Kick, and Peter Biczo, "Psychometric Properties of the Beck Depression Inventory," in *Present, Past and Future of Psychiatry*, vol. 1, eds. A. Beigel, J.J. Lopez Ibor, Jr., and J.A. Costa e Silva (Singapore: World Scientific Publishing, 1994), 247–49.

33. Weatherill et al., "Is Maternal Depression Related to Side of Infant Holding?"

34. Peter de Château, Hertha Holmberg, and Jan Winberg, "Left-Side Preference in Holding and Carrying Newborn Infants I: Mothers Holding and Carrying During the First Week Life," *Acta Paediatrica: Nurturing the Child* 67, no. 2 (March 1978): 169–75.

35. Mi Li, Hongpei Xu, and Shengfu Lu, "Neural Basis of Depression Related to a Dominant Right Hemisphere: A Resting-State fMRI Study," *Behavioural Neurology* (2018): 1–10, https://downloads.hindawi.com/journals/bn/2018/5024520.pdf.

36. Lea-Ann Pileggi et al., "Cradling Bias Is Absent in Children with Autism Spectrum Disorders," *Journal of Child and Adolescent Mental Health* 25, no. 1 (2013): 55–60.

37. Gianluca Malatesta et al., "The Role of Ethnic Prejudice in the Modulation of Cradling Lateralization," *Journal of Nonverbal Behavior* 45 (2021): 187–205.

38. J.T. Manning and J. Denman, "Lateral Cradling Preferences in Humans (*Homo sapiens*): Similarities Within Families," *Journal of Comparative Psychology* 108, no. 3 (September 1994): 262–65.

39. Michelle Tomaszycki et al., "Maternal Cradling and Infant Nipple Preferences in Rhesus Monkeys (*Macaca mulatta*)," *Developmental Psychobiology* 32, no. 4 (May 1998): 305–12.

40. Takeshi Hatta and Motoko Koike, "Left-Hand Preference in Frightened Mother Monkeys in Taking Up Their Babies," *Neuropsychologia* 29, no. 2 (1991): 207–09.

41. Ichirou Tanaka, "Change of Nipple Preference Between Successive Offspring in Japanese Macaques," *American Journal of Primatology* 18, no. 4 (1989): 321–25, https://doi.org/10.1002/ajp.1350180406.

42. Karina Karenina, Andrey Giljov, and Yegor Malashichev, "Lateralization of Mother-Infant Interactions in Wild Horses," *Behavioural Processes* 148 (March 2018): 49–55, https://doi.org/10.1016/j.beproc.2018.01.010.

43. Andrey Giljov, Karina Karenina, and Yegor Malashichev, "Facing Each Other: Mammal Mothers and Infants Prefer the Position Favouring Right Hemisphere Processing," *Biology Letters* 14, no. 1 (January 2018): 20170707.

44. Karenina, Giljov, and Malashichev, "Lateralization of Mother-Infant Interactions in Wild Horses."

45. Karina Karenina, Andrey Giljov, Shermin de Silva, and Yegor Malashichev, "Social Lateralization in Wild Asian Elephants: Visual Preferences of Mothers and Offspring," *Behavioral Ecology and Sociobiology* 72, no. 21 (2018).

46. Stephen E. Palmer, Karen B. Schloss, and Jonathan Sammartino, "Visual Aesthetics and Human Preference," *Annual Review of Psychology* 64, no. 1 (January 2013): 77–107.

CHAPTER 6 POSING BIASES: PUTTING THE BEST CHEEK FORWARD

1. Erna Bombeck, *When You Look Like Your Passport Photo, It's Time to Go Home* (New York: Random House Value Publishing, 1993).

2. I.C. McManus and N.K. Humphrey, "Turning the Left Cheek," *Nature* 243 (June 1973): 271–72.

3. Charles Darwin, *The Expression of the Emotions in Man and Animals* (London: John Murray, 1872).

4. Charles Darwin, *On the Origin of Species by Means of Natural Selection, or Preservation of Favoured Races in the Struggle for Life* (London: John Murray, 1859).

5. Paul Ekman and Wallace V. Friesen, *Pictures of Facial Affect* (Berkeley, CA: Consulting Psychologists Press, 1976).

6. Paul Ekman, "An Argument for Basic Emotions," *Cognition and Emotion* 6, nos. 3–4 (1992): 169–200.

7. Joan C. Borod, Elissa Koff, and Betsy White, "Facial Asymmetry in Posed and Spontaneous Expressions of Emotion," *Brain and Cognition* 2, no. 2 (April 1983): 165–75.

8. Borod et al., "Emotional and Non-Emotional Facial Behaviour in Patients with Unilateral Brain Damage," *Journal of Neurology, Neurosurgery, and Psychiatry* 51, no. 6, (1988): 826–32, https://doi.org/10.1136/jnnp.51.6.826.

9. Ruth Campbell, "Asymmetries in Interpreting and Expressing a Posed Facial Expression," *Cortex: A Journal Devoted to the Study of the Nervous System and Behavior* 14, no. 3 (1978): 327–42.

10. Harold A. Sackeim, Ruben C. Gur, and Marcel Saucy, "Emotions Are Expressed More Intensely on the Left Side of the Face," *Science* 202, no. 4366 (October 27, 1978): 434–36.

11. Martin Skinner and Brian Mullen, "Facial Asymmetry in Emotional Expression: A Meta-Analysis of Research," *British Journal of Social Psychology* 30, no. 2 (1991): 113–24.

12. Patten, *Neurological Differential Diagnosis*.

13. McManus and Humphrey, "Turning the Left Cheek."

14. Carolyn J. Mebert and George F. Michel, "Handedness in Artists," in *Neuropsychology of Left-Handedness*, ed. Jeannine Herron (Toronto: Academic Press, 1980), 273–79.

15. Mary A. Peterson and Gillian Rhodes, eds., *Perception of Faces, Objects, and Scenes: Analytic and Holistic Processes* (New York: Oxford University Press, 2003).

16. James W. Tanaka and Martha J. Farah, "Parts and Wholes in Face Recognition," *The Quarterly Journal of Experimental Psychology* 46, no. 2 (June 1993): 225–45.

17. Annukka K. Lindell, "The Silent Social/Emotional Signals in Left and Right Cheek Poses: A Literature Review," *Laterality* 18, no. 5 (2013): 612–24.

18. Miyuki Yamamoto et al., "Accelerated Recognition of Left Oblique Views of Faces," *Experimental Brain Research* 161, no. 1 (February 2005): 27–33.

19. Nicola Bruno, Marco Bertamini, and Federica Protti, "Selfie and the City: A World-Wide, Large, and Ecologically Valid Database Reveals a Two-Pronged Side Bias in Naïve Self-Portraits," *PLoS ONE* 10 no. 4 (April 27, 2015): e0124999, https://doi.org/10.1371/journal.pone.0124999.

20. Annukka K. Lindell, "Capturing Their Best Side? Did the Advent of the Camera Influence the Orientation Artists Chose to Paint and Draw in Their Self-Portraits?" *Laterality* 18, no. 3 (2013): 319–28.

21. Annukka K. Lindell, Tenenbaum, and Aznar, "Left Cheek Bias for Emotion Perception, but Not Expression, Is Established in Children Aged 3–7 Years," *Laterality* 22, no. 1 (2017): 17–30, http://dx.doi.org/10.1080 /1357650X.2015.1108328.

22. Bruno, Bertamini, and Protti, "Selfie and the City."

23. Michael E.R. Nicholls, Danielle Clode, Stephen J. Wood, and Amanda J. Wood, "Laterality of Expression in Portraiture: Putting Your Best Cheek Forward," *Proceedings of the Royal Society B: Biological Sciences* 266, no. 1428 (September 1999): 1517–22, https://doi.org/10.1098/rspb.1999.0809.

24. Carel ten Cate, "Posing as Professor: Laterality in Posing Orientation for Portraits of Scientists," *Journal of Nonverbal Behavior* 26, no. 3 (2002): 175–92.

25. Nicholls, Clode, Wood, and Wood, "Laterality of Expression in Portraiture."

26. McManus, "Half a Century of Handedness Research."

27. McManus, "Half a Century of Handedness Research."

28. Matia Okubo and Takato Oyama, "Do You Know Your Best Side? Awareness of Lateral Posing Asymmetries," *Laterality* (2021): 1–15, https:// doi.org/10.1080/1357650X.2021.1938105.

29. Owen Churches et al., "Facing Up to Stereotypes: Surgeons and Physicians Are No Different in Their Emotional Expressiveness," *Laterality* 19, no. 5 (2014): 585–90.

30. Churches et al., "Facing Up to Stereotypes."

31. Churches et al., "Facing Up to Stereotypes."

32. Acosta, Williamson, and Heilman, "Which Cheek Did Jesus Turn?"

33. Lealani Mae Y. Acosta, John B. Williamson, and Kenneth B. Heilman, "Which Cheek Did the Resurrected Jesus Turn?" *Journal of Religion and Health* 54, no. 3 (June 2015): 1091–98, http://dx.doi.org/10.1007/s10943-014-9945-9.

34. Acosta, Williamson, and Heilman, "Which Cheek Did the Resurrected Jesus Turn?"

35. Kari N. Duerksen, Trista E. Friedrich, and Lorin J. Elias, "Did Buddha Turn the Other Cheek Too? A Comparison of Posing Biases Between Jesus and Buddha," *Laterality* 21, nos. 4–6 (July-November 2016): 633–42, http://dx.doi.org/10.1080/1357650X.2015.1087554.

36. Duerksen, Friedrich, and Elias, "Did Buddha Turn the Other Cheek Too?"

37. Nicole A. Thomas, Jennifer A. Burkitt, and Deborah M. Saucier, "Photographer Preference or Image Purpose? An Investigation of Posing Bias in Mammalian and Non-Mammalian Species," *Laterality* 11, no. 4 (July 2006): 350–54.

CHAPTER 7 LIGHTING BIASES: DO WE HAVE THE RIGHT LIGHTING?

1. Mark Twain, *Mark Twain at Your Fingertips: A Book of Quotations*, ed. Caroline Thomas Harnsberger (Mineola, NY: Dover, 2009).

2. Ian Christopher McManus, Joseph Buckman, and Euan Woolley, "Is Light in Pictures Presumed to Come from the Left Side?" *Perception* 33, no. 12 (2004): 1421–36.

3. Kevin S. Berbaum, Todd Bever, and Chan Sup Chung, "Light Source Position in the Perception of Object Shape," *Perception* 12, no. 4 (1983): 411–16.

4. Jennifer Sun and Pietro Perona, "Where Is the Sun?" *Nature: Neuroscience* 1, no. 3 (1998): 183–84.

5. Sun and Perona, "Where Is the Sun?"

6. McManus, Buckman, and Woolley, "Is Light in Pictures Presumed to Come from the Left Side?"

7. David A. McDine, Ian J. Livingston, Nicole A. Thomas, Lorin J. Elias, "Lateral Biases in Lighting of Abstract Artwork," *Laterality* 16, no. 3 (May 2011): 268–79.

8. Kobayashi et al., "Natural Preference in Luminosity for Frame Composition," *NeuroReport* 18, no. 11 (2007): 1137–40.

9. Sun and Perona, "Where Is the Sun?"

10. Pascal Mamassian and Ross Goutcher, "Prior Knowledge on the Illumination Position," *Cognition* 81, no. 1 (September 2001): B1–9.

11. McManus, Buckman, and Woolley, "Is Light in Pictures Presumed to Come from the Left Side?"

12. Austen K. Smith, Izabela Szelest, Trista E. Friedrich, and Lorin J. Elias, "Native Reading Direction Influences Lateral Biases in the Perception of Shape from Shading," *Laterality* 20, no. 4 (2015): 418–33.

13. Mark E. McCourt, Barbara Blakeslee, and Ganesh Padmanabhan, "Lighting Direction and Visual Field Modulate Perceived Intensity of Illumination," *Frontiers in Psychology* 4 no. 983 (December 2013): 1–6.

14. Jennifer R. Sedgewick, Bradley Weiers, Aaron Stewart, and Lorin J. Elias, "The Thinker: Opposing Directionality of Lighting Bias Within Sculptural Artwork," *Frontiers in Human Neuroscience* 9, no. 251 (May 2015): 1–8.

15. Austen K. Smith, Jennifer R. Sedgewick, Bradley Weiers, and Lorin J. Elias, "Is There an Artistry to Lighting? The Complexity of Illuminating Three-Dimensional Artworks," *Psychology of Aesthetics, Creativity, and the Arts* 15, no. 1 (2021): 20–27.

16. Smith, Szelest, Friedrich, and Elias, "Native Reading Direction Influences Lateral Biases in the Perception of Shape from Shading."

17. Bridget Andrews, Daniela Aisenberg, Giovanni d'Avossa, and Ayelet Sapir, "Cross-Cultural Effects on the Assumed Light Source Direction: Evidence from English and Hebrew Readers," *Journal of Vision* 13, no. 13 (November 2013): 1–7.

18. Nicole A. Thomas, Jennifer A. Burkitt, Regan A. Patrick, and Lorin J. Elias, "The Lighter Side of Advertising: Investigating Posing and Lighting Biases," *Laterality* 13, no. 6 (November 2008): 504–13.

CHAPTER 8 SIDE EFFECTS IN ART, AESTHETICS, AND ARCHITECTURE

1. Harold J. McWhinnie, "Is Psychology Relevant to Aesthetics?" *Proceedings of the Annual Convention of the American Psychological Association* 6, part 1 (1971): 419–20.

2. George Dickie, "Is Psychology Relevant to Aesthetics?" *The Philosophical Review* 71, no. 3 (July 1962): 285–302.

3. Annukka K. Lindell and Julia Mueller, "Can Science Account for Taste?

Psychological Insights into Art Appreciation," *Journal of Cognitive Psychology* 23, no. 4 (2011): 453–75.

4.   Rolf Reber, "Art in Its Experience: Can Empirical Psychology Help Assess Artistic Value?" *Leonardo* 41, no. 4 (August 2008): 367–72.

5.   Lindell and Mueller, "Can Science Account for Taste?"

6.   Hermann Weyl, *Symmetry* (Princeton, NJ: Princeton University Press, 1952).

7.   Ian Christopher McManus, "Symmetry and Asymmetry in Aesthetics and the Arts," *European Review* 13, supplement 2 (2005): 157–80.

8.   John P. Anton, "Plotinus' Refutation of Beauty as Symmetry," *The Journal of Aesthetics and Art Criticism* 23, no. 2 (Winter 1964): 233–37.

9.   McManus, "Symmetry and Asymmetry in Aesthetics and the Arts."

10.   Mercedes Gaffron, "Some New Dimensions in the Phenomenal Analysis of Visual Experience," *Journal of Personality* 24, no. 3 (1956): 285–307.

11.   Heinrich Wölfflin, "Über das rechts und links im Bilde," in *Gedanken zur Kunstgeschichte: Gedrucktes und Ungedrucktes*, 3rd ed., ed. Heinrich Wölfflin (Basel, Switzerland: Schwabe & Co., 1941), 82–90.

12.   Charles G. Gross and Marc H. Bornstein, "Left and Right in Science and Art," *Leonardo* 11, no. 1 (Winter 1978): 29–38.

13.   Gross and Bornstein, "Left and Right in Science and Art."

14.   Samy Rima et al., "Asymmetry of Pictorial Space: A Cultural Phenomenon," *Journal of Vision* 19, no. 4 (April 2019): 1–6.

15.   Wölfflin, "Über das rechts und links im Bilde."

16.   Gross, "Left and Right in Science and Art."

17.   Lindell and Mueller, "Can Science Account for Taste?"

18.   Rudolf Arnheim, *Art and Visual Perception: A Psychology of the Creative Eye* (Berkeley, CA: University of California Press, 1974).

19.   Gaffron, "Some New Dimensions in the Phenomenal Analysis of Visual Experience."

20.   Wölfflin, "Über das rechts und links im Bilde."

21.   Gross, "Left and Right in Science and Art."

22.   Carmen Pérez González, "Lateral Organisation in Nineteenth-Century Studio Photographs Is Influenced by the Direction of Writing: A Comparison of Iranian and Spanish Photographs," *Laterality* 17, no. 5 (September 2012): 515–32.

23.   Sobh Chahboun et al., "Reading and Writing Direction Effects on the Aesthetic Perception of Photographs," *Laterality* 22, no. 3 (May 2017): 313–39.

24. Trista E. Friedrich, Victoria L. Harms, and Lorin J. Elias, "Dynamic Stimuli: Accentuating Aesthetic Preference Biases," *Laterality* 19, no. 5 (2014): 549–59.

25. Trista E. Friedrich and Lorin J. Elias, "The Write Bias: The Influence of Native Writing Direction on Aesthetic Preference Biases," *Psychology of Aesthetics, Creativity, and the Arts* 10, no. 2 (2016): 128–33.

26. Friedrich, Harms, and Elias, "Dynamic Stimuli."

27. Friedrich and Elias, "The Write Bias."

28. Marilyn Freimuth and Seymour Wapner, "The Influence of Lateral Organization on the Evaluation of Paintings," *British Journal of Psychology* 70, no. 2 (1979): 211–18.

29. Thomas M. Nelson and Gregory A. MacDonald, "Lateral Organization, Perceived Depth, and Title Preference in Pictures," *Perceptual and Motor Skills* 33, no. 3, part 1 (1971): 983–86.

30. Barry T. Jensen, "Reading Habits and Left-Right Orientation in Profile Drawings by Japanese Children," *The American Journal of Psychology* 65, no. 2 (April 1952): 306–07.

31. Barry T. Jensen, "Left-Right Orientation in Profile Drawing," *The American Journal of Psychology* 65, no. 1 (January 1952): 80–83.

32. Sylvie Chokron, Seta Kazandjian, and Maria De Agostini, "Effects of Reading Direction on Visuospatial Organization: A Critical Review," in *Quod Erat Demonstrandum: From Herodotus' Ethnographic Journeys to Cross-Cultural Research: Proceedings from the 18th International Congress of the International Association for Cross-Cultural Psychology*, eds. Aikaterini Gari and Kostas Mylonas (Athens, Greece: Pedio Books Publishing, 2009), 107–14.

33. Sümeyra Tosun and Jyotsna Vaid, "What Affects Facing Direction in Human Facial Profile Drawing? A Meta-Analytic Inquiry," *Perception* 43, no. 12 (December 2014): 1377–92.

34. Alexander G. Page, Ian Christopher McManus, Carmen Pérez González, and Sobh Chahboun, "Is Beauty in the Hand of the Writer? Influences of Aesthetic Preferences Through Script Directions, Cultural, and Neurological Factors: A Literature Review," *Frontiers in Psychology* 8 (August 2017): 1–10.

35. Anjan Chatterjee, Lynn M. Maher, and Kenneth M. Heilman, "Spatial Characteristics of Thematic Role Representation," *Neuropsychologia* 33, no. 5 (1995): 643–48.

36. Anjan Chatterjee, M. Helen Southwood, and David Basilico, "Verbs, Events and Spatial Representations," *Neuropsychologia* 37, no. 4 (1999): 395–402.

37. Anne Maass, Caterina Suitner, Xenia Favaretto, and Marina Cignacchi, "Groups in Space: Stereotypes and the Spatial Agency Bias," *Journal of Experimental Social Psychology* 45, no. 3 (May 2009): 496–504, http:// dx.doi.org/10.1016/j.jesp.2009.01.004.

38. Caterina Suitner and Anne Maass, "Spatial Agency Bias: Representing People in Space," *Advances in Experimental Social Psychology* 53 (January 2016): 245–301.

39. Maass, Suitner, Favaretto, and Cignacchi, "Groups in Space."

40. Maass, Suitner, Favaretto, and Cignacchi, "Groups in Space."

41. Caterina Suitner, Anne Maass, and Lucia Ronconi, "From Spatial to Social Asymmetry: Spontaneous and Conditioned Associations of Gender and Space," *Psychology of Women Quarterly* 41, no. 1 (March 2017): 46–64.

42. Anne Maass, Caterina Suitner, and Faris Nadhmi, "What Drives the Spatial Agency Bias? An Italian-Malagasy-Arabic Comparison Study," *Journal of Experimental Psychology: General* 143, no. 3 (2014): 991–96.

43. Mara Mazzurega, Maria Paola Paladino, Claudia Bonfiglioli, and Susanna Timeo, "Not the Right Profile: Women Facing Rightward Elicit Responses in Defence of Gender Stereotypes," *Psicologia sociale* 14, no. 1 (2019): 57–72.

44. Dilip Kondepudi and Daniel J. Durand, "Chiral Asymmetry in Spiral Galaxies?" *Chirality* 13, no. 7 (July 2001): 351–56, https://doi.org/10.1002/ chir.1044.

45. Robert Couzin, "The Handedness of Historiated Spiral Columns," *Laterality* 22, no. 5 (November 2017): 1–31.

46. Heinz Luschey, *Rechts und Links: Untersuchungen über Bewegungsrichtung, Seitenordnung und Höhenordnung als Elemente der antiken Bildsprache* (Tübingen, Germany: Wasmuth, 2002).

47. Couzin, "The Handedness of Historiated Spiral Columns."

CHAPTER 9 GESTURES: LEFTOVER BEHAVIOURAL FOSSILS

1. Jana M. Iverson, Heather L. Tencer, Jill Lany, and Susan Goldin-Meadow, "The Relation Between Gesture and Speech in Congenitally Blind and Sighted Language-Learners," *Journal of Nonverbal Behavior* 24, no. 2 (2000): 105–30.

2. Sotaro Kita, "Cross-Cultural Variation of Speech-Accompanying Gesture: A Review," *Language and Cognitive Processes* 24, no. 2 (2009): 145–67.

3. Elias and Saucier, *Neuropsychology*.

4. Gordon W. Hewes et al., "Primate Communication and the Gestural Origin of Language [and Comments and Reply]," *Current Anthropology* 14, nos. 1–2 (February-April 1973): 5–24.

5. Michael C. Corballis, *The Lopsided Ape: Evolution of the Generative Mind* (New York: Oxford University Press, 1991).

6. Merlin Donald, "Preconditions for the Evolution of Protolanguages," in *The Descent of Mind: Psychological Perspectives on Hominid Evolution*, eds. Michael C. Corballis and Stephen E.G. Lea (New York: Oxford University Press, 1999), 138–54.

7. Doreen Kimura, "Manual Activity During Speaking: I. Right-Handers," *Neuropsychologia* 11, no. 1 (1973): 45–50.

8. Doreen Kimura, "Manual Activity During Speaking: II. Left-Handers," *Neuropsychologia* 11, no. 1 (1973): 51–55.

9. John Thomas Dalby, David Gibson, Vittorio Grossi, and Richard D. Schneider, "Lateralized Hand Gesture During Speech," *Journal of Motor Behavior* 12, no. 4 (1980): 292–97.

10. Deborah M. Saucier and Lorin J. Elias, "Lateral and Sex Differences in Manual Gesture During Conversation," *Laterality* 6, no. 3 (July 2001): 239–45.

11. Lorin J. Harris, "Hand Preference in Gestures and Signs in the Deaf and Hearing: Some Notes on Early Evidence and Theory," *Brain and Cognition* 10, no. 2 (July 1989): 189–219.

12. Sotaro Kita and Hedda Lausberg, "Generation of Co-Speech Gestures Based on Spatial Imagery from the Right-Hemisphere: Evidence from Split-Brain Patients," *Cortex* 44, no. 2 (2008): 131–39.

13. Kita and Lausberg, "Generation of Co-Speech Gestures Based on Spatial Imagery from the Right-Hemisphere."

14. Elias and Saucier, *Neuropsychology.*

15. Paraskevi Argyriou, Christine Mohr, and Sotaro Kita, "Hand Matters: Left-Hand Gestures Enhance Metaphor Explanation," *Journal of Experimental Psychology: Learning Memory and Cognition* 43, no. 6 (2017): 874–86.

16. Argyriou et al., "Hand Matters: Left-Hand Gestures Enhance Metaphor Explanation."

17. Gordon W. Hewes, "Primate Communication and the Gestural Origin of Language," *Current Anthropology* 33, no. 1, supplement (February 1992): 65–84.

18. Gordon W. Hewes, "An Explicit Formulation of the Relationship Between Tool-Using, Tool-Making, and the Emergence of Language," *Visible Language* 7, no. 2 (Spring 1973): 101–27.

19. Michael C. Corballis, "Did Language Evolve Before Speech?" in *The Evolution of Human Language: Biolinguistic Perspectives*, eds. Richard K. Lawson, Viviane Déprez, and Hiroko Yamakido (Cambridge: Cambridge University Press, 2010), 115–23.

20. Giacomo Rizzolatti and Michael A. Arbib, "Language Within Our Grasp," *Trends in Neurosciences* 21, no. 5 (May 1998): 188–94.

21. Giacomo Rizzolatti, Leonardo Fogassi, and Vittorio Gallese, "Neurophysiological Mechanisms Underlying the Understanding and Imitation of Action," *Nature Reviews Neuroscience* 2, no. 9 (September 2001): 661–70.

22. Corballis, "Did Language Evolve Before Speech?"

CHAPTER 10 TURNING BIASES: THINGS THAT GO BUMP ON THE RIGHT

1. Gaspard Gustave Coriolis, "Sur les équations du mouvement relatif des systèmes de corps," in *Journal de l'École Royale Polytechnique, Cahier XXIV, Tome XV* (Paris: Bachelier, 1835), 144–54.

2. Theo Gerkema and Louis Gostiaux, "A Brief History of the Coriolis Force," *Europhysics News* 43, no. 2 (March 2012): 14–17.

3. P.Y. Hennion and R. Mollard, "An Assessment of the Deflecting Effect on Human Movement Due to the Coriolis Inertial Forces in a Space Vehicle," *Journal of Biomechanics* 26, no. 1 (January 1993): 85–90.

4. Ingrid A.P. Ververs, Johanna I.P. de Vries, Hermann P. van Geijn, and Brian Hopkins, "Prenatal Head Position from 12–38 Weeks. I.

Developmental Aspects," *Early Human Development* 39, no. 2 (October 1994): 83–91.

5. *Zoolander*, directed by Ben Stiller (Paramount, 2001), DVD.

6. Tino Stöckel and Christian Vater, "Hand Preference Patterns in Expert Basketball Players: Interrelations Between Basketball-Specific and Everyday Life Behavior," *Human Movement Science* 38 (December 2014): 143–51.

7. Eve Golomer et al., "The Influence of Classical Dance Training on Preferred Supporting Leg and Whole Body Turning Bias," *Laterality* 14, no. 2 (September 2009): 165–77.

8. Dora Stratou, *The Greek Dances: Our Living Link with Antiquity* (Athens: A. Klissiounis, 1966).

9. Catherine Augé and Yvonne Paire, *L'engagement corporel dans les danses traditionnelles de France métropolitaine* (Paris: Ministère de la Culture, 2006).

10. S.F. Ali, K.J. Kordsmeier, and B. Gough, "Drug-Induced Circling Preference in Rats," *Molecular Neurobiology* 11, nos. 1–3 (August-December 1995): 145–54, https://doi.org/10.1007/BF02740691.

11. A.A. Schaeffer, "Spiral Movement in Man," *Journal of Morphology* 45, no. 1 (1928): 293–398, http://doi.wiley.com/10.1002/jmor.1050450110.

12. Edward S. Robinson, "The Psychology of Public Education," *American Journal of Public Health* 23, no. 2 (February 1933): 123–28.

13. Robinson, "The Psychology of Public Education," 125.

14. Peter G. Hepper, Glenda R. McCartney, and E. Alyson Shannon, "Lateralised Behaviour in First Trimester Human Foetuses," *Neuropsychologia* 36, no. 6 (June 1998): 531–34.

15. Hepper, McCartney, and Shannon, "Lateralised Behaviour in First Trimester Human Foetuses."

16. B. Hopkins, W. Lems, Beatrice Janssen, and George Butterworth, "Postural and Motor Asymmetries in Newlyborns," *Human Neurobiology* 6, no. 3 (1987): 153–56.

17. Sonya Dunsirn et al., "Defining the Nature and Implications of Head Turn Preference in the Preterm Infant," *Early Human Development* 96 (May 2016): 53–60, http://dx.doi.org/10.1016/j.earlhumdev.2016.02.002.

18. Yukio Konishi, Haruki Mikawa, and Junko Suzuki, "Asymmetrical Head-Turning of Preterm Infants: Some Effects on Later Postural and Functional Lateralities," *Developmental Medicine & Child Neurology* 28, no. 4 (1986): 450–57, http://doi.wiley.com/10.1111/j.1469-8749.1986.tb14282.x.

19. Arnold Gesell, "The Tonic Neck Reflex in the Human Infant: Morphogenetic and Clinical Significance," *The Journal of Pediatrics* 13, no. 4 (1938): 455–64.

20. John Reiser, Albert Yonas, and Karin Wikner, "Radial Localization of Odors by Human Newborns," *Child Development* 47 (1976): 856–59.

21. Jane Coryell, and George F. Michel, "How Supine Postural Preferences of Infants Can Contribute Toward the Development of Handedness," *Infant Behavior & Development* 1 (1978): 245–57.

22. H. Stefan Bracha, David J. Seitz, John Otemaa, and Stanley D. Glick, "Rotational Movement (Circling) in Normal Humans: Sex Difference and Relationship to Hand, Foot, and Eye Preference," *Brain Research* 411, no. 2 (1987): 231–35.

23. Bracha, Seitz, Otemaa, and Glick, "Rotational Movement (Circling) in Normal Humans."

24. Schaeffer, "Spiral Movement in Man."

25. Larissa A. Mead and Elizabeth Hampson, "Turning Bias in Humans Is Influenced by Phase of the Menstrual Cycle," *Hormones and Behavior* 31, no. 1 (1997): 65–74.

26. Schaeffer, "Spiral Movement in Man."

27. Richard Morris, "Developments of a Water-Maze Procedure for Studying Spatial Learning in the Rat," *Journal of Neuroscience Methods* 11, no. 1 (1984): 47–60.

28. Peng Yuan, Ana M. Daugherty, and Naftali Raz, "Turning Bias in Virtual Spatial Navigation: Age-Related Differences and Neuroanatomical Correlates," *Biological Psychology* 96 (February 2014): 8–19, http://dx.doi.org/10.1016/j.biopsycho.2013.10.009.

29. Pablo Covarrubias, Ofelia Citlalli López-Jiménez, and Ángel Andrés Jiménez Ortiz, "Turning Behavior in Humans: The Role of Speed of Locomotion," *Conductal* 2, no. 2 (2014): 39–50.

30. Matthieu Lenoir, Sophie van Overschelde, Myriam De Rycke, and Emilienne Musch, "Intrinsic and Extrinsic Factors of Turning Preferences in Humans," *Neuroscience Letters* 393, nos. 2–3 (2006): 179–83.

31. M. Yanki Yazgan, James F. Leckman, and Bruce E. Wexler, "A Direct Observational Measure of Whole Body Turning Bias," *Cortex* 32, no. 1 (1996): 173–76, http://dx.doi.org/10.1016/S0010-9452(96)80025-6.

32. John L. Bradshaw and Judy A. Bradshaw, "Rotational and Turning

Tendencies in Humans: An Analog of Lateral Biases in Rats?" *The International Journal of Neuroscience* 39, nos. 3–4 (1988): 229–32.

33. Stratou, "The Greek Dances."

34. M.J.D. Taylor, S.C. Strike, and P. Dabnichki, "Turning Bias and Lateral Dominance in a Sample of Able-Bodied and Amputee Participants," *Laterality* 12, no. 1 (2006): 50–63.

35. Sarah B. Wallwork et al., "Left/Right Neck Rotation Judgments Are Affected by Age, Gender, Handedness and Image Rotation," *Manual Therapy* 18, no. 3 (2013): 225–30, http://dx.doi.org/10.1016/j. math.2012.10.006.

36. Emel Güneş and Erhan Nalçaci, "Directional Preferences in Turning Behavior of Girls and Boys," *Perceptual and Motor Skills* 102, no. 2 (2007): 352–57.

37. H.D. Day and Kaaren C. Day, "Directional Preferences in the Rotational Play Behaviors of Young Children," *Developmental Psychobiology* 30, no. 3 (1997): 213–23.

38. Day and Day, "Directional Preferences in the Rotational Play Behaviors of Young Children."

39. Christine Mohr, H. Stefan Bracha, T. Landis, and Peter Brugger, "Opposite Turning Behavior in Right-Handers and Non-Right-Handers Suggests a Link Between Handedness and Cerebral Dopamine Asymmetries," *Behavioral Neuroscience* 117, no. 6 (2003): 1448–52.

40. Christine Mohr et al., "Human Side Preferences in Three Different Whole-Body Movement Tasks," *Behavioural Brain Research* 151, nos. 1–2 (2004): 321–26.

41. Jan Stochl and Tim Croudace, "Predictors of Human Rotation," *Laterality* 18, no. 3 (2013): 265–81.

42. Oliver H. Turnbull and Peter McGeorge, "Lateral Bumping: A Normal-Subject Analog to the Behaviour of Patients with Hemispatial Neglect?" *Brain and Cognition* 37, no. 1 (1998): 31–33.

43. Dawn Bowers and Kenneth M. Heilman, "Pseudoneglect: Effects of Hemispace on a Tactile Line Bisection Task," *Neuropsychologia* 18, nos. 4–5 (January 1980): 491–98.

44. Michael E.R. Nicholls, Andrea Loftus, Kerstin Mayer, and Jason B. Mattingley, "Things That Go Bump in the Right: The Effect of Unimanual Activity on Rightward Collisions," *Neuropsychologia* 45, no. 5 (March 14, 2007): 1122–26.

45. Michael E.R. Nicholls et al., "A Hit-and-Miss Investigation of Asymmetries in Wheelchair Navigation," *Attention Perception & Psychophysics* 72, no. 6 (August 2010): 1576–90.

46. Nicholls et al., "A Hit-and-Miss Investigation of Asymmetries in Wheelchair Navigation."

47. Robinson, "The Psychology of Public Education," 128.

CHAPTER 11 SEATING BIASES: 2B OR NOT 2B?

1. Paul R. Farnsworth, "Seat Preference in the Classroom," *The Journal of Social Psychology* 4, no. 3 (1933): 373–76, https://doi.org/10.1080/00224545.1933.9919330.

2. L.L. Morton and J.R. Kershner, "Hemisphere Asymmetries, Spelling Ability, and Classroom Seating in Fourth Graders," *Brain and Cognition* 6, no. 1 (1987): 101–11.

3. Robert Sommer, "Classroom Ecology," *The Journal of Applied Behavioral Science* 3, no. 4 (1967): 489–502, https://doi.org/10.1177/002188636700300404.

4. Paul Bakan, "The Eyes Have It," *Psychology Today* 4 (1971): 64–69.

5. Raquel E. Gur, Ruben C. Gur, and Brachia Marshalek, "Classroom Seating and Functional Brain Asymmetry," *Journal of Educational Psychology* 67, no. 1 (1975): 151–53.

6. Gur, Gur, and Marshalek, "Classroom Seating and Functional Brain Asymmetry."

7. Ruben C. Gur, Harold A. Sackeim, and Raquel E. Gur, "Classroom Seating and Psychopathology: Some Initial Data," *Journal of Abnormal Psychology* 85, no. 1 (1976): 122–24.

8. Elias and Saucier, *Neuropsychology*.

9. Morton and Kershner, "Hemisphere Asymmetries, Spelling Ability, and Classroom Seating in Fourth Graders."

10. Victoria L. Harms, Lisa J.O. Poon, Austen K. Smith, and Lorin J. Elias, "Take Your Seats: Leftward Asymmetry in Classroom Seating Choice," *Frontiers in Human Neuroscience* 9, no. 457 (2015).

11. Harms, Poon, Smith, and Elias, "Take Your Seats."

12. George B. Karev, "Cinema Seating in Right, Mixed and Left Handers," *Cortex* 36, no. 5 (2000): 747–52.

13. Peter Weyers, Annette Milnik, Clarissa Müller, and Paul Pauli, "How to

Choose a Seat in Theatres: Always Sit on the Right Side?" *Laterality* 11, no. 2 (March 2006): 181–93, https://doi.org/10.1080/13576500500430711.

14. Matia Okubo, "Right Movies on the Right Seat: Laterality and Seat Choice," *Applied Cognitive Psychology* 24, no. 1 (2010): 90–99.

15. Victoria Lynn Harms, Miriam Reese, and Lorin J. Elias, "Lateral Bias in Theatre-Seat Choice," *Laterality* 19, no. 1 (2014): 1–11.

16. Oliver Smith, "Most Popular Aircraft Seat Revealed," *Telegraph*, April 11, 2013, telegraph.co.uk/travel/news/Most-popular-aircraft-seat-revealed.

17. Michael E.R. Nicholls, Nicole A. Thomas, and Tobias Loetscher, "An Online Means of Testing Asymmetries in Seating Preference Reveals a Bias for Airplanes and Theaters," *Human Factors* 55, no. 4 (2013): 725–31.

18. Stephen Darling, Dario Cancemi, and Sergio Della Sala, "Fly on the Right: Lateral Preferences When Choosing Aircraft Seats," *Laterality* 23, no. 5 (2018): 610–24.

CHAPTER 12 SPORTS: COMPETING THE RIGHT WAY

1. Coren and Porac, "Fifty Centuries of Right-Handedness."

2. Eero Vuoksimaa, Markku Koskenvuo, Richard J. Rose, and Jaakko Kaprio, "Origins of Handedness: A Nationwide Study of 30 161 Adults," *Neuropsychologia* 47, no. 5 (2009): 1294–1301.

3. Michel Raymond and Dominique Pontier, "Is there Geographical Variation in Human Handedness?" *Laterality* 9, no. 1 (January 2004): 35–51.

4. Steven Pinker, *The Better Angels of Our Nature: Why Violence Has Declined* (New York: Viking, 2011), 802.

5. Napoleon A. Chagnon, *Yaṇomamö: The Fierce People* (New York: Holt, Rinehart & Winston, 1983).

6. Napoleon A. Chagnon, "Life Histories, Blood Revenge, and Warfare in a Tribal Population," *Science* 239, no. 4843: 985–92.

7. Michel Raymond, Dominique Pontier, Anne-Béatrice Dufour, and Anders Pape Møller, "Frequency-Dependent Maintenance of Left-Handedness in Humans," *Proceedings of the Royal Society B: Biological Sciences* 263, no. 1377 (1996): 1627–33.

8. Thomas V. Pollet, Gert Stulp, and Ton G.G. Groothuis, "Born to Win? Testing the Fighting Hypothesis in Realistic Fights: Left-Handedness in the Ultimate Fighting Championship," *Animal Behaviour* 86, no. 4 (2013):

839–43, http://dx.doi.org/10.1016/j.anbehav.2013.07.026.

9. Roger N. Shepard and Jacqueline Metzler, "Mental Rotation of Three-Dimensional Objects," *Science* 171, no. 3972 (1971): 701–03.

10. Lorin J. Harris, "In Fencing, What Gives Left-Handers the Edge? Views from the Present and the Distant Past," *Laterality* 15, nos. 1–2 (2010): 15–55.

11. Guy Azémar and J.F. Stein, "Surreprésentation des gauchers, en fonction de l'arme, dans l'elite mondiale de l'escrime," paper presented at the Congrès International de la Société Française de Psychologie du Sport in Poitiers, France, in 1994.

12. Olympics Statistics, "Edoardo Mangiarotti," databaseolympics.com, databaseolympics.com/players/playerpage.htm?ilkid=MANGIEDO01.

13. Harris, "In Fencing, What Gives Left-Handers the Edge?"

14. Olympics Statistics, "Edoardo Mangiarotti."

15. Thomas Richardson and R. Tucker Gilman, "Left-Handedness Is Associated with Greater Fighting Success in Humans," *Science Reports* 9, no. 15402 (2019).

16. "How to Score a Fight," BoxRec, http://boxrec.com/media/index.php/ How_to_Score_a_Fight.

17. Richardson and Gilman, "Left-Handedness Is Associated with Greater Fighting Success in Humans."

18. Mehmet Akif Ziyagil, Recep Gursoy, Şenol Dane, and Ramazan Yuksel, "Left-Handed Wrestlers Are More Successful," *Perceptual and Motor Skills* 111, no. 1 (2011): 65–70.

19. Yunus Emre Cingoz et al., "Research on the Relation Between Hand Preference and Success in Karate and Taekwondo Sports with Regards to Gender," *Advances in Physical Education* 8, no. 3 (2018): 308–20.

20. Recep Gursoy et al., "The Examination of the Relationship Between Left-Handedness and Success in Elite Female Archers," *Advances in Physical Education* 7, no. 4 (2017): 367–76.

21. Pollet, Stulp, and Groothuis, "Born to Win?"

22. Florian Loffing and Norbert Hagemann, "Pushing Through Evolution? Incidence and Fight Records of Left-Oriented Fighters in Professional Boxing History," *Laterality* 20, no. 3 (2015): 270–86.

23. Robert Brooks, Luc F. Bussière, Michael D. Jennions, and John Hunt, "Sinister Strategies Succeed at the Cricket World Cup," *Proceedings of the Royal Society B: Biological Science* 271, supplement 3 (2004): S64–S66.

24. Wei-Chun Wang et al., "Preferences in Athletes: Insights from a Database of 1770 Male Athletes," *American Journal of Sports Science* 6, no. 1 (2018): 20–25.

25. Florian Loffing, Norbert Hagemann, Jörg Schorer, and Joseph Baker, "Skilled Players' and Novices' Difficulty Anticipating Left- vs. Right-Handed Opponents' Action Intentions Varies Across Different Points in Time," *Human Movement Science* 40 (2015): 410–21, http://dx.doi.org/10.1016/j.humov.2015.01.018.

26. Florian Loffing, Jörg Schorer, Norbert Hagemann, and Joseph Baker, "On the Advantage of Being Left-Handed in Volleyball: Further Evidence of the Specificity of Skilled Visual Perception," *Attention, Perception, and Psychophysics* 74, no. 2 (2012): 446–53.

27. Francois Fagan, Martin Haugh, and Hal Cooper, "The Advantage of Lefties in One-on-One Sports," *Journal of Quantitative Analysis in Sports* 15, no. 1 (2019): 1–25.

28. Belo Petro and Attila Szabo, "The Impact of Laterality on Soccer Performance," *Strength and Conditioning Journal* 38, no. 5 (October 2016): 66–74.

29. Hassane Zouhal et al., "Laterality Influences Agility Performance in Elite Soccer Players," *Frontiers in Physiology* 9, no. 807 (June 2018): 1–8.

30. Benjamin B. Moore et al., "Laterality Frequency, Team Familiarity, and Game Experience Affect Kicking-Foot Identification in Australian Football Players," *International Journal of Sports Science and Coaching* 12, no. 3 (2017): 351–58.

31. Josu Barrenetxea-Garcia, Jon Torres-Unda, Izaro Esain, and Susana M. Gil, "Relative Age Effect and Left-Handedness in World Class Water Polo Male and Female Players," *Laterality* 24, no. 3 (2019): 259–73.

32. Florian Loffing, Norbert Hagemann, and Bernd Strauss, "Left-Handedness in Professional and Amateur Tennis," *PLoS ONE* 7, no. 11 (2012): 1–8.

33. Barrenetxea-Garcia, Torres-Unda, Esain, and Gil, "Relative Age Effect and Left-Handedness in World Class Water Polo Male and Female Players."

34. Moore et al., "Laterality Frequency, Team Familiarity, and Game Experience Affect Kicking-Foot Identification in Australian Football Players."

35. Florian Loffing, Norbert Hagemann, and Bernd Strauss, "Automated Processes in Tennis: Do Left-Handed Players Benefit from the Tactical

Preferences of Their Opponents?" *Journal of Sports Sciences* 28, no. 4 (2010): 435–43.

36. Loffing, Hagemann, and Strauss, "Automated Processes in Tennis."

37. Loffing, Hagemann, and Strauss, "Automated Processes in Tennis."

38. Alex Bryson, Bernd Frick, and Rob Simmons, "The Returns to Scarce Talent: Footedness and Player Remuneration in European Soccer," *Journal of Sports Economics* 14, no. 6 (2013): 606–28.

39. Lorin J. Elias, M.P. Bryden, and M.B. Bulman-Fleming, "Footedness Is a Better Predictor Than Is Handedness of Emotional Lateralization," *Neuropsychologia* 36, no. 1 (1998): 37–43.

40. Lorin J. Elias, M.B. Bulman-Fleming, and Murray J. Guylee, "Complementarity of Cerebral Function Among Individuals with Atypical Laterality Profiles," *Brain and Cognition* 40, no. 1 (1999): 112–15.

41. Elias and Saucier, *Neuropsychology.*

42. Jan Verbeek et al., "Laterality Related to the Successive Selection of Dutch National Youth Soccer Players," *Journal of Sports Sciences* 35, no. 22 (2017): 2220–2224.

43. Lawler and Lawler, "Left-Handedness in Professional Basketball."

44. Florian Loffing, "Left-Handedness and Time Pressure in Elite Interactive Ball Games," *Biology Letters* 13, no. 11 (2017).

45. Loffing, Schorer, Hagemann, and Baker, "On the Advantage of Being Left-Handed in Volleyball."

46. Michael E.R. Nicholls, Tobias Loetscher, and Maxwell Rademacher, "Miss to the Right: The Effect of Attentional Asymmetries on Goal-Kicking," *PLoS ONE* 5, no. 8 (2010): 1–6.

47. Ross Roberts and Oliver H. Turnbull, "Putts That Get Missed on the Right: Investigating Lateralized Attentional Biases and the Nature of Putting Errors in Golf," *Journal of Sports Sciences* 28, no. 4 (2010): 369–74.

48. J.P. Coudereau, Nils Guéguen, M. Pratte, and Eliana Sampaio, "Tactile Precision in Right-Handed Archery Experts with Visual Disabilities: A Pseudoneglect Effect?" *Laterality* 12, no. 2 (2006): 170–80.

49. Martin Dechant et al., "In-Game and Out-of-Game Social Anxiety Influences Player Motivations, Activities, and Experiences in MMORPGs," in *Proceedings of the 2020 CHI Conference on Human Factors in Computing Systems* (New York: Association for Computing Machinery, 2020), 1–14, https://dl.acm.org/doi/fullHtml/10.1145/3313831.3376734.

50. Andrew J. Roebuck et al., "Competitive Action Video Game Players
     Display Rightward Error Bias During On-Line Video Game Play,"
     Laterality 23, no. 5 (2018): 505–16.
51. Anne Maass, Damiano Pagani, and Emanuela Berta, "How Beautiful Is the
     Goal and How Violent Is the Fistfight? Spatial Bias in the Interpretation of
     Human Behavior," *Social Cognition* 25, no. 6 (2007): 833–52.

Image Credits

93	ID 167077829 by Anatolii63@Dreamstime.com.
95	ID 150169703 by Anatolii63@Dreamstime.com.
96	ID 184446808 by Iuliia Selina@Dreamstime.com.
100	Lorin J. Elias.
101	Adapted from ID 216057999 by Archangel80889@Dreamstime.com.
103	Adapted from ID 156155797 by Maxim Ivasiuk@Dreamstime.com.
104	Wikimedia Commons, public domain works of art.
106	Naoharu Koayashi.
108	Lorin J. Elias.
111	ID 113948281 by Martina1802@Dreamstime.com.
115 (top)	ID 53410192 by Mili387@Dreamstime.com.
115 (bottom)	ID 20716662 by Hara Sahani@Dreamstime.com.
116	Wikimedia Commons, public domain works of art.
117	Wikimedia Commons, public domain works of art.
119	Adapted from ID 120732458 by Peter Hermes Furian@Dreamstime.com.
120	Jean-Honoré Fragonard, *Pastoral Scene*, public domain artwork.
121	Adapted from ID 7228587 by Pavel Losevsky@Dreamstime.com.
125	Lorin J. Elias, adapted from Marra Mazzurega.
126 (right)	Adapted from ID 19574713 by Stelya@Dreamstime.com.
126 (left)	Adapted from Wikimedia Commons.
127	ID 58216390 by Yuliia Yakovyna@Dreamstime.com.
133	Wikimedia Commons, public domain work.
134	Lorin J. Elias.
135	Mileva Elias.
146	Wikimedia Commons, public domain work.
148	Lorin J. Elias.
154	Adapted from ID 141758584 by Ustyna Shevhcuk@Dreamstime.com.
159	Lorin J. Elias, adapted from an image by Victoria Harms.
162	Lorin J. Elias, adapted from an image by Victoria Harms.
164	Adapted from ID 90708094 by Nitinut380@Dreamstime.com.
170	Wikimedia Commons, public domain work.
173	ID 26873780 by Fabio Brocchi@Dreamstime.com.
176	ID 1934104 by Nicholas Rjabow@Dreamstime.com.
178	pngkit.com.
180	Adapted from ID 7310661 by Guilu@Dreamstime.com.

Index

thyroid disease, 22
Trajan's Column, 127
trans-tibial, 147
triadic, 22
Twain, Mark, 100
twins, 19, 23–25

ulcerative colitis, 23
ultrasound, 24
Urdu, 59, 109, 122, 148

van Gogh, Vincent, 86, 118
Vermeer, 85
    Art of Painting, The, 86
vertebrate, 4
Vietnamese, 16

volleyball, 172, *176–77*, 179

walrus, 78
water polo, 172
Watson, Thomas, 15–16
Western art, 45, 119, 128
Wölfflin, Heinrich, 117, 118
wrong-sided surgeries, 43
Wurtz, Felix
    Children's Book, The, 67

yamiin, 47

Zatorre, Robert, 38
zurdo, 47

About the Author

Lorin Elias is an award-winning professor of psychology and vice-dean academic at the University of Saskatchewan. He completed his Ph.D. in behavioural neuroscience at the University of Waterloo and has been studying left-brain right-brain differences for over twenty-five years. His federally funded research program has resulted in over seventy research papers, appearing in journals such as *Neurosurgery, Neuropsychologia, Cortex, Brain and Cognition, Laterality, Cognitive Brain Research,* and *Behavioural Neuroscience.* These studies have been featured in articles in popular newspapers and magazines around the world, such as *Cosmopolitan, Wired, Maxim,* the *New York Post,* and the *New Zealand Herald.* Beyond his accomplishments in research, Dr. Elias is also a celebrated professor of psychology. He has won several teaching awards at the University of Saskatchewan, including the Master Teacher Award. Dr. Elias has earned a reputation for making seemingly complicated topics and esoteric material accessible, fascinating, and even fun. He has co-authored a college-level neuropsychology textbook and an adaptation of an introductory psychology textbook. Dr. Elias offers a variety of courses from introductory psychology up to graduate-level studies in cognitive neuroscience. He also writes for fun, including reviews of high-end audio equipment.